Lyotropic Liquid Crystals

and the Structure of Biomembranes

Stig Friberg, EDITOR

The Swedish Institute for Surface Chemistry

A symposium based on

The Fifth International

Liquid Crystal Conference

Proceedings.

Stockholm, Sweden,

June 17–21, 1974.

ADVANCES IN CHEMISTRY SERIES **152**

AMERICAN CHEMICAL SOCIETY

WASHINGTON, D. C. 1976

Library of Congress CIP Data

Lyotropic liquid crystals and the structure of biomembranes.
(Advances in chemistry series; 152. ISSN 0065-2393)

Includes biographical references and index.

1. Liquid crystals—Congresses. 2. Membranes (Biology)—Congresses.
I. Friberg, Stig, 1930- II. Title. III. Series.

QD1.A355 no. 152 (QD923) 540'.8s
(548'.9) 76-18704
ISBN 0-8412-0269-9 ADCSAJ 152 1-156 (1976)

PRINTED IN THE UNITED STATES OF AMERICA

Advances in Chemistry Series

Robert F. Gould, *Editor*

FOREWORD

Advances in Chemistry Series was founded in 1949 by the American Chemical Society as an outlet for symposia and collections of data in special areas of topical interest that could not be accommodated in the Society's journals. It provides a medium for symposia that would otherwise be fragmented, their papers distributed among several journals or not published at all. Papers are refereed critically according to ACS editorial standards and receive the careful attention and processing characteristic of ACS publications. Papers published in Advances in Chemistry Series are original contributions not published elsewhere in whole or major part and include reports of research as well as reviews since symposia may embrace both types of presentation.

CONTENTS

PREFACE

The importance of association structures of amphiphilic molecules for interfacial phenomena has been realized in the last ten years with the rapid progress of knowledge concerning the structure of biomembranes and the discovery of the pronounced influence of surfactant association structures on the properties of disperse systems.

During the discussions at the Fifth International Liquid Crystal Conference in Stockholm it was found that long range order is the common factor in these systems. The discussions gave the impetus to combine the present knowledge of the structure of liquid crystals with that of pertinent interfacial phenomena in a volume to serve as reference book for researchers—such as scientists working in the fields of pharmacy, foods or cosmetics—with an interest in both biostructures and disperse systems.

This volume covers the structural relations between thermotropic and lyotropic liquid crystals (Chapters 1 and 2) and compares them with the micellar systems (Chapter 3). The interfacial aspects and the accompanying stability problems are covered in Chapters 5 and 6. The molecular dynamics in liquid crystals, the importance of water structure and of counter-ion binding for their stability are three essential factors for long range order systems, which are treated in Chapters 7, 8, and 9. The final chapter by E. J. Ambrose illustrates the change of order in a biological system under malignant conditions.

It is a pleasure to express my gratitude to the participants in the Stockholm conference for their interest and enthusiasm, and to the authors of the volume, who generously shared their knowledge in an exciting and rapidly developing field of science.

Stockholm
May 1976

STIG FRIBERG

INTRODUCTION

The Planning and Steering Committee for International Liquid Crystal Conferences recommended that the proceedings of the Fifth International Liquid Crystal Conference should be published in two sources. Most of the papers on thermotropic liquid crystals will be published in *Journal de Physique,* and those dealing with lyotropic systems will be collected as a group and published in this hardback volume. The Committee asked Stig Friberg to serve as editor of this volume. We are pleased that this volume is a part of the American Chemical Society's *Advances in Chemistry Series.*

The papers appearing in this volume represent the common research areas encountered in lyotropic systems. An effort is made to include articles connecting the lyotropic liquid crystals with biological structures characterized by similar long-range order phenomena. The papers contained in this book should serve not only researchers in the field who want to expand their knowledge of lyotropic liquid crystals but also those who are starting in the field.

GLENN H. BROWN, CHAIRMAN
Planning & Steering Committee

January 1975

1

Generic Relationships between Non-Amphiphilic and Amphiphilic Mesophases of the "Fused" Type

Relationship of Cubic Mesophases ("Plastic Crystals") Formed by Non-Amphiphilic Globular Molecules to Cubic Mesophases of the Amphiphilic Series

G. W. GRAY

Department of Chemistry, The University, Hull, HU6 7RX, England

P. A. WINSOR

Shell Research, Ltd., Thornton Research Centre, Chester, CH1 3SH, England

A generic relationship exists between lamellar mesophases (smectics A and C), formed by non-amphiphilic mesogens with lath-like molecules and "fused" lamellar mesophases, G (with or without solvents), formed by amphiphilic mesogens. A second generic relationship occurs between non-amphiphilic nematic mesophases and two-dimensionally hexagonal "middle" mesophases (M_1, M_2), formed by fibrous amphiphilic micelles. A third generic relationship is now proposed between three-dimensionally periodic cubic "rotational" mesophases ("plastic crystals"), formed by non-amphiphilic mesogens with globular molecules, and cubic mesophases (S_{1c}, V_1, V_2), formed by globular amphiphilic micelles. Finally it is suggested that the cubic mesophase, "smectic D," formed by a few non-amphiphilic mesogens with lath-like molecules, is a rotational mesophase based on globular groupings (of parallel molecules) that arise in the transition between smectics A and C.

The purpose of this account is to point out certain generic relationships between the mesophases of the non-amphiphilic series—nematic, smectic, and plastic crystal—and the fused mesophases—middle (M_1 and M_2), smectic (G), and cubic (S_{1c}, V_1, and V_2)—of the amphiphilic series.

The cubic amphiphilic mesophases (S_{1c}, V_1, and V_2) from their interposition in the succession of mesophases S_{1c}, M_1, V_1, G, V_2, and M_2, have generally been termed "liquid crystalline" like the optically anisotropic amphiphilic mesophases M_1, G, and M_2. The cubic mesophases formed by non-amphiphilic globular molecules have however usually been termed "plastic crystals." This nomenclature has obscured the fact that these "plastic crystals" are fundamentally liquid crystals rather than solid cyrstals and bear a relationship to the optically anisotropic non-amphiphilic smectic and nematic liquid crystals similar to that born by the amphiphilic cubic mesophases to the optically anisotropic "neat" (G) and middle (M_1 and M_2) liquid crystalline phases.

Mesophases of Nematic Type

The intermicellar equilibrium which, according to the R theory (*1, 2, 3*) is responsible for the succession of micellar amphiphilic solution phases, both amorphous and liquid crystalline, is shown in Figure 1.

Analogously to the formation of the nematic mesophase in the non-amphiphilic series by the parallel arrangement of lath-like molecules without other long range positional or orientational order (Figure 2a), in the amphiphilic series the middle mesophases M_1 and M_2 are formed by the parallel arrangement of cylindrical fibrous micelles. These micelles, because of their indefinite extension, are arranged in a two-dimensional hexagonal lattice (Figure 3) which confers a high viscosity on these M phases that is not found with non-amphiphilic nematic phases.

Mesophases of Smectic Type

In the non-amphiphilic series the smectic A mesophase is constituted by an extension of the parallel molecular ordering in the nematic phase so that the parallel lath-like molecules are grouped in parallel indefinitely extended sheets (Figure 2b). In the amphiphilic series, the smectic mesophase G (Figure 4) is formed by the parallel arrangement of indefinitely extended lamellar "sandwich" micelles in which the amphiphilic molecules have a polar head-to-polar head position and hydrocarbon tail-to-hydrocarbon tail arrangement with solvents, if present, partitioned between the $\overline{O}$, $\overline{C}$, and $\overline{W}$ regions (Ref. *3*, p. 273) according to their polarity.

In the non-amphiphilic smectic A mesophase (Figure 2b), the parallel, fairly rigid, lath-like molecules are grouped with orientational disorder (at right angles to their long axes) and end-to-end so that the molecules lie statistically normal to the sheets, constituting an optically uniaxial

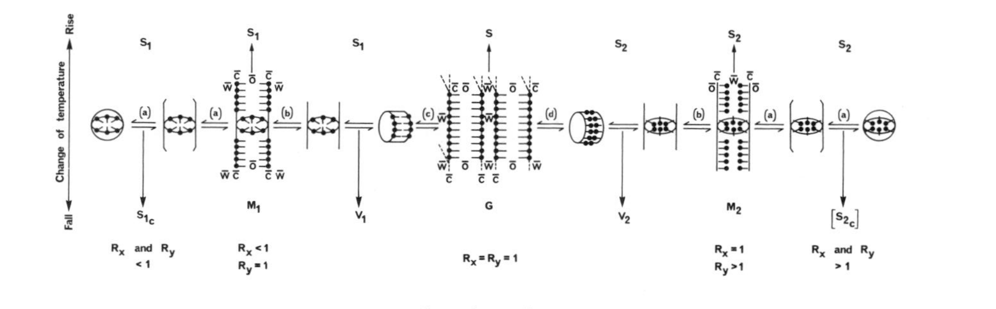

Figure 1. Nature of the succession of amphiphilic mesophases of the fused type in relation to the underlying micellar equilibria (1, 2).

$$R = \frac{\text{tendency of } \overline{C} \text{ to be convex towards } \overline{O}}{\text{tendency of } \overline{C} \text{ to be convex towards } \overline{W}}$$

When the polar groups are ordered on the micellar face in two-dimensional nematic order, x *and* y *correspond to the directions on the micellar face corresponding to the minimum and maximum values of* R *respectively* (3).

Legend:

S_1 and S_2 = *rotational and positional disorder*
S_{1c} and S_{2c} = *rotational disorder; units in three-dimensional cubic lattice;* S_{2c} *is not yet definitely identified*
V_1 and V_2 = *rotational disorder, units in three dimensional cubic lattice*
M_1 and M_2 = *two-dimensional hexagonal lattice of indefinitely extended parallel cylindrical micelles*
G = *lamellar lattice of indefinitely extended parallel lamellar micelles*
$\overline{W}$ = *aqueous zone*
$\overline{C}$ = *amphiphilic zone*
$\overline{O}$ = *hydrocarbon zone*
(a) = *breakdown of long rods caused by tendency to develop convexities longitudinally*
(b) = *breakdown of indefinite extension of long rods caused by tendency to develop planar regions circumferentially*
(c) = *breakdown of indefinite lamellae caused by tendency to develop convexities towards* $\overline{W}$
(d) = *breakdown of lamellae caused by development of convexities towards* $\overline{O}$

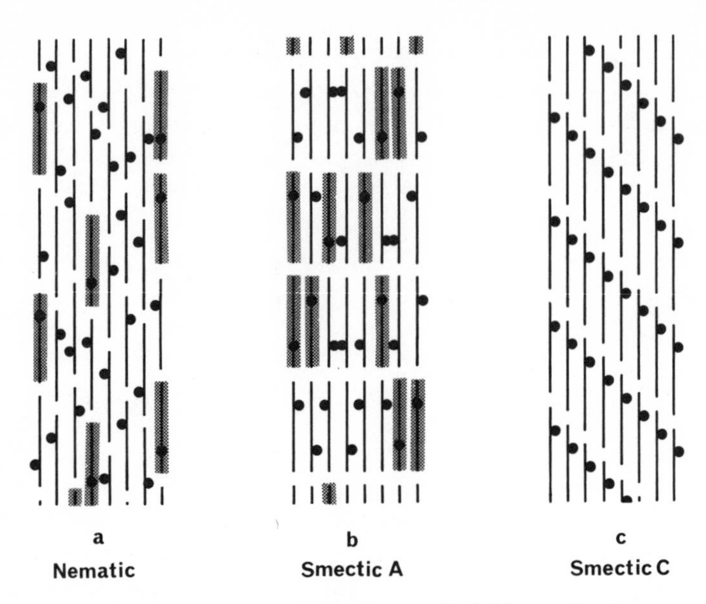

Figure 2. Types of arrangement of lath-like molecules in certain non-amphiphilic mesophases

The black disk represents any distinguishing feature on one face of the lath-like molecule which is shown as a line when viewed side on and as a dotted area when viewed face on. Each section (a), (b), and (c) represents a cross-section through an idealized arrangement with the molecules aligned longitudinally in the plane of the drawing, parallel to a plane surface that lies at right angles to the plane of the drawing and parallel to the margin. This cross-section will repeat indefinitely above and below the plane of the drawing.

(a) Mean longitudinal parallelism of the molecules is the sole regular feature of the nematic arrangement, the molecules being otherwise disordered in the plane of the drawing and in the plane at right angles to the direction of mean parallelism. In the latter plane, the cross-sections of the molecules are in complete two-dimensional isotropic disorder. Although the molecules in the drawing are shown as equidistant and parallel, this is intended to represent a statistical mean equidistance and parallelism on either side of which there will be considerable deviations.

(b) In addition to mean longitudinal parallelism, each molecule conditions the mean positions of its neighbors so that there is a minimizing of the mean distance between their ends or centers. This gives rise to layers perpendicular to the direction of mean parallelism. There is no additional regular feature within this smectic A molecular arrangement.

(c) Additional regularities of molecular arrangement result in the mean relative positions of minimum energy for neighbors lying not with their midpoints at a minimized mean distance of separation but with particular non-corresponding points at this minimized distance. This results in layers inclined to the direction of mean parallelism of the molecules. There is no evidence for such uniform polar order as shown in this section which has been drawn like this solely for simplicity. The average situation through many layers of a smectic C may be such that there is no resultant polarity of the order.

phase. In the amphiphilic G mesophase (Figure 4), the flexible hydrocarbon chains are not grouped like parallel rods but are arranged as in a liquid hydrocarbon, subject to the restriction of the attached polar groups into parallel sheets. According to the packing of the polar groups—*i.e.*, according to the interfacial area per polar group on the $\overline{W}$ face of the micelle, the thickness of the liquid hydrocarbon region, $\overline{O}$, of the micelle may vary with concentration and/or temperature, and in certain circumstances may be considerably less than twice the fully extended hydrocarbon chain length. Nonetheless, there is no preferred mean direction of inclination of C–C bonds with respect to the lamellar planes so that like smectic A the G phase is optically uniaxial.

In certain smectic phases (not considered in detail here) of the non-amphiphilic series, there is a degree of temperature-dependent ordering of the parallel molecules greater than that occurring in smectic A phases. For example, the lath-like molecules may possess some degree of orientational ordering at right angles to their long axes—*i.e.*, a two-dimensional nematic ordering—and may group other than with geometrically corresponding points at closest approach distance. This may result in the planes of the sheets of molecules not lying at right angles to the molecular long axes—smectic C (Figure 2c).

In the smectic phases of the amphiphilic series, although the hydrocarbon chains themselves have a liquid arrangement, a degree of ordering of the polar ends of the molecules ($-CH_2X$) on the polar face of the micelles is possible. This ordering would be expected to vary with tem-

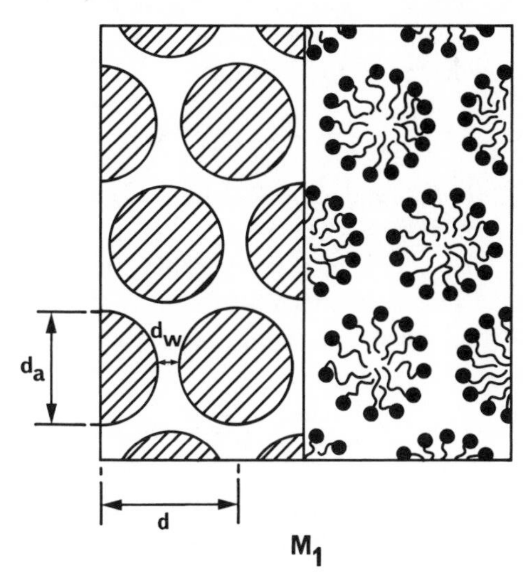

Figure 3. Two-dimensional, hexagonal structure of the middle mesophase M_1 in relation to the measured x-ray long spacing, d

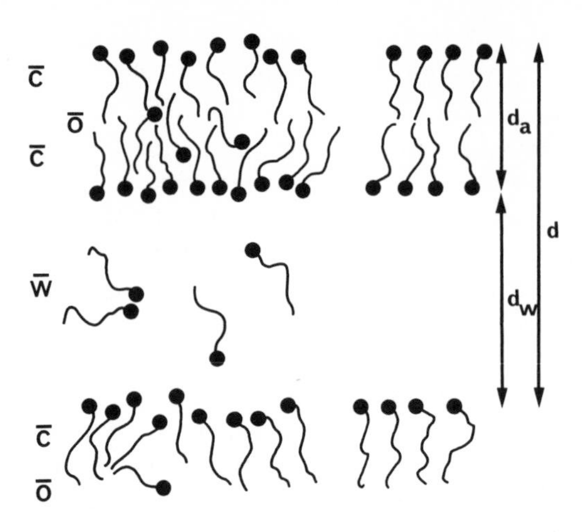

Figure 4. Diagrammatic representation of the structure of a G mesophase in a binary amphiphile/water system.

Left of drawing: $\overline{C}$ layer contains most of amphiphile, but both amphiphile and water are distributed throughout the $\overline{C}$, $\overline{O}$, and $\overline{W}$ regions so that their respective activities are statistically uniform throughout. Right of drawing: working approximation usually used in x-ray studies. The $\overline{C}$ layer contains only and entirely all of the amphiphile present.

perature and water content and may account for certain amphiphilic smectic polymorphs (*e.g.*, neat soap phase, soap boiler's neat phase G, Ekwall's phase B) reported in the literature.

Mesophases of Cubic or "Plastic Crystal" Type

Amphiphilic Cubic Mesophases; Viscous Isotropic Phases. In Figure 1, at positions where a high concentration of globular micellar forms might be expected, the cubic "viscous isotropic" mesophases S_{1c}, V_1, and V_2 are encountered. This is further illustrated by Figure 5. These mesophases, because of their inclusion within the general sequence of amphiphilic mesophases:

$$S_{1c}, M_1, V_1, G, V_2, M_2$$

have usually been recognized as liquid crystals. However, in early work, Luzzati pointed out their three-dimensional crystalline characteristics (*4, 5, 6*).

Non-Amphiphilic Cubic Mesophases; Plastic Crystals. The cubic "plastic crystals" of the non-amphiphilic series, to which the amphiphilic

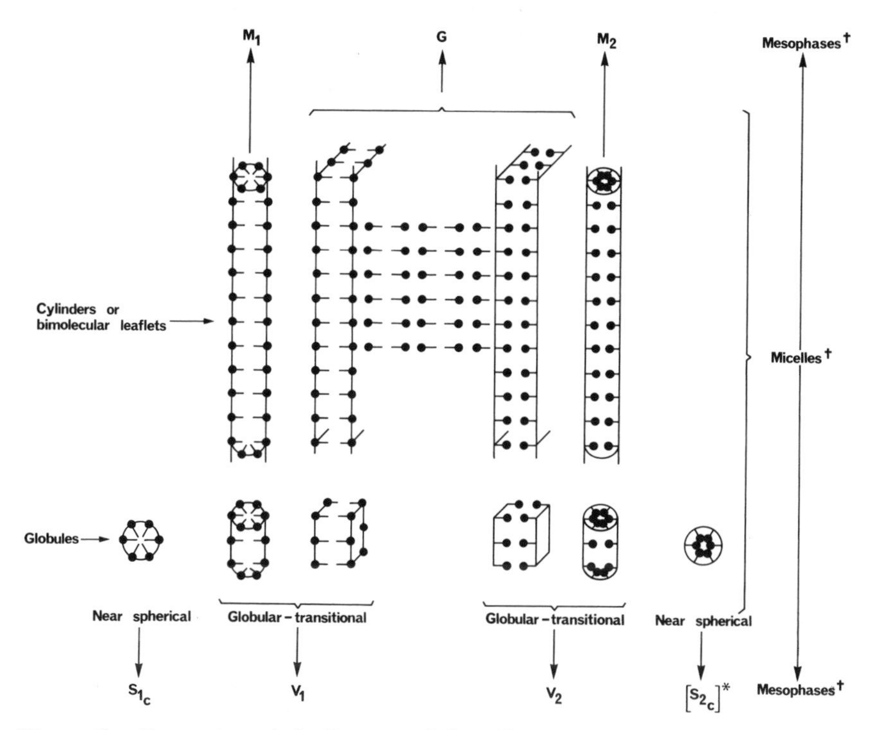

Figure 5. Formation of the hexagonal, lamellar, and cubic amphiphilic mesophases by the interactions between fibrous, lamellar, and globular micelles, respectively

S_{1c}, V_1, V_2, and S_{2c}	= *optically isotropic cubic mesophases;*
M_1 and M_2	= *anisotropic, optically birefringent, nematic type mesophases —parallel, indefinitely extended cylinders in two dimensional hexagonal array*
G	= *anisotropic, optically birefringent, lamellar smectic A type mesophase—parallel, indefinitely extended equidistant leaflets*
+	= *Mesophases formed by the operation of intermicellar forces at sufficiently close intermicellar distance and separation, from micellar forms, fluid in character, organized by the operation of intramicellar forces dependent on composition and temperature*

cubic viscous isotropic mesophases are generically analogous, have usually been regarded as truly crystalline rather than as liquid crystalline. However, their "rotational" mesomorphous character was clearly recognized by Timmermanns, their discoverer (7), and he (8) illustrated their generic relationship to the optically anisotropic nematic and smectic liquid crystals as in Figure 6. This relationship is further illustrated by the thermodynamic data in Table I (9), in which the relatively high heats of transition from solid crystal to mesophase and the relatively low heats of transition from mesophase to amorphous liquid appear with both the anisotropic and the cubic classes of mesophase.

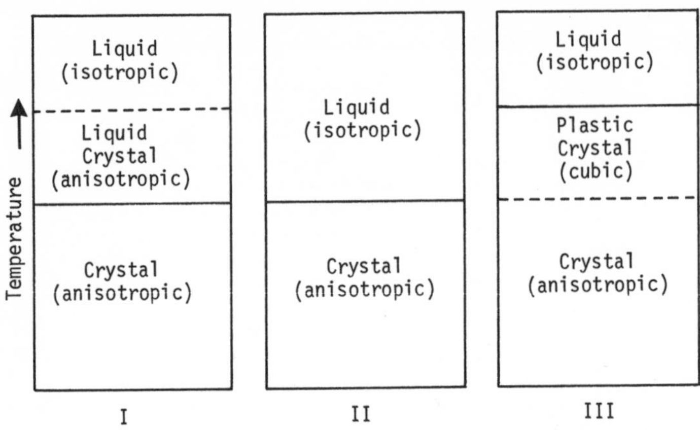

Figure 6. Phases for systems which give: I, crystal, anisotropic liquid, and isotropic liquid; II, crystal and isotropic liquid; III, crystal, cubic plastic crystal, and isotropic liquid (8)

Whereas the parallel structural units of nematic and smectic liquid crystals are rod-like and lamellar, respectively, in plastic (liquid) crystals the units are either globular molecules (*see* Table I) or globular groups of molecules in the non-amphiphilic series, or they are globular micelles in the amphiphilic series. Many types of "globular" organic molecules are known which give rise to plastic crystals. These molecules need not be strictly spherical but may be prolate ($Cl_3C \cdot CCl_3$) or oblate spheroids (cyclohexane) or even strawberry shaped ($(CH_3)_3C \cdot CH_2 \cdot CH_3$). The important point is that within the plastic crystal lattice they undergo fairly free thermal rotational motions and thus acquire effectively near

Table I. Comparative Thermodynamic Data for Some Compounds Giving Transitions from Solid Crystal to Amorphous Liquid Either with or without an Intermediate Mesophase

Compound	*Mesophase*	T_c (°C)[a]	H_c (cal/g)[b]	T_a (°C)[c]	H_a (cal/g)[d]
n-Hexane	none	−95	36.0	$=T_c$	$=H_c$
Cyclohexane	plastic	−87	19.0	6.5	7.5
cis-1,2-Dimethylcyclohexane	plastic	−100.7	16.7	−50	3.5
trans-1,2-Dimethylcyclohexane	none	−88.2	22.4	$=T_c$	$=H_c$
Ethyl *p*-azoxybenzoate	smectic	114	13.8	122.6	3.5
p-Azoxyanisole	nematic	118	27.0	135	0.53

[a] T_c = temperature of breakdown of the solid crystal lattice.
[b] H_c = latent heat of transition at T_c.
[c] T_a = temperature of transition to amorphous liquid. For non-mesogens, $T_a = T_c$.
[d] H_a = latent heat of transition at T_a. For non-mesogens, $H_a = H_c$.

Table II. Crystalline Forms of Some Non-Amphiphilic Plastic Crystals (*23, 24*) and of Some Amphiphilic Cubic Mesophases

Material	*Cubic Plastic Crystal (Classification of Cubic Lattice)*	*Solid Crystals at Lower Temperatures; Crystal Type*
CBr_4	Pa3	monoclinic
$CCl_3 \cdot CCl_3$	Im3m	triclinic
Cyclopentane	(hexagonal)	
Cyclohexane	fcc	
S_{1c} dodecyltrimethylammonium chloride/water	Pm3n	
V_1 dimethyldodecylamine oxide/water	fcc	
V_1, V_2 various, anhydrous or with water	1_a3d, Pm3n	

spherical symmetry (*9*). Such restriction of rotation as occurs determines the type of cubic lattice adopted (Table II). With completely free rotation, a face-centered cubic lattice would be expected.

One feature of the non-amphiphilic cubic mesophases is that they frequently show mutual miscibility even when constituted from dissimilar molecules. Such miscibility, which contrasts with the immiscibility between dissimilarly constituted solid crystals, is also found between nematic mesophases, between corresponding smectic polymorphs, and, of course, between amorphous liquids. This miscibility is important in its implication that in the cubic mesophases of the amphiphilic series there could well be an equilibrium of related globular micellar forms (Figures 1 and 5) rather than a single clearly defined form.

An important characteristic of the cubic mesophases, either non-amphiphilic or amphiphilic, is that because of the fairly free thermal rotational motions of their constituent units, they typically give high resolution NMR spectra. In this respect they behave like amorphous liquids and quite differently from conventional solids or from mesophases such as M_1, M_2, or G in which rotation of the units is more severely restricted.

Shortcomings of Alternative Model Structures for Cubic Phases S_{1c}, V_1, and V_2

A number of continuous network, jointed-rod models for the structures of the S_{1c}, V_1, and V_2 phases have been proposed by Luzzati and his collaborators (*10, 11, 12*) on the basis of x-ray diffraction measurements. In these models, the individual rods are close to isodimensional and thus represent globular micelles, but these are pictured, not as rotating at the lattice points but as jointed into continuous interpenetrating networks so as to confer rigidity on the structure. Perhaps the main objection to these models is that, in contrast to rotational plastic crystal structures (*13*), they would not lead one to expect these phases to show their characteristic high resolution NMR spectra but to behave similarly to the middle and neat mesophases which do not give such spectra.

It seems to the present authors that to account for the x-ray measurements it may be possible to replace Luzzati's jointed rods by rotational globular micelles geometrically arranged in a manner related to that postulated for the jointed rods. The stability and rigidity of the phase would then be attributed to lattice forces (van der Waals attractions and repulsions, forces opposing the interpenetration of electrical double layers) similar to those that confer stability and some rigidity on the micellar lattices of M_1, M_2, and G mesophases. According to Luzzati (*14*) and also according to the R theory (*1, 2, 3, 15*), the micelles themselves are almost devoid of internal rigidity. It will therefore be the intermicellar forces—*i.e.*, the lattice forces—rather than forces from jointing or close packing (*16*) which confer the viscosities and stabilities of the mesophases

Conditions for the Formation of Amphiphilic Mesophases

From this point of view, the following conditions must be satisfied for an amphiphilic mesophase to form:

(1) Micelles of appropriate conformation—*i.e.*, size and shape—must be present.

(2) The micelles of this size and shape (*i.e.*, in sufficient close proximity) must be concentrated sufficiently to result in the formation of the mesophase lattice at the prevailing temperature. At some higher temperature, this lattice will break down to give the amorphous liquid by the disintegrating effect of the increased thermal motion.

Formation of the S_{1c} Phase in Binary Aqueous Alkyltrimethylammonium Halide Solutions. These considerations are well illustrated by the formation of the S_{1c} phase in aqueous alkyltrimethylammonium halide solutions (*17*) (Figure 7). This phase is apparently composed of S_1 micelles—probably on balance prolate—arranged in a primitive, cubic lattice and rotating fairly freely at the lattice points. The lattice is formed by dodecyl- and tetradecyltrimethylammonium chlorides but not by the hexadecyl or octadecyl chlorides nor by any of the corresponding bromides. This may be expressed as follows.

In Figure 1, within a series of homologous amphiphiles, a given stage in the micellar progression is reached at a lower concentration the higher the hydrocarbon chain length of the amphiphile is. With the hexadecyl and octadecyltrimethylammonium chlorides conversion of globular (S_1) to fibrous (M_1) micelles apparently occurs at too low a concentration—*i.e.*, at too great an intermicellar distance—for S_{1c} lattice to form. Replace-

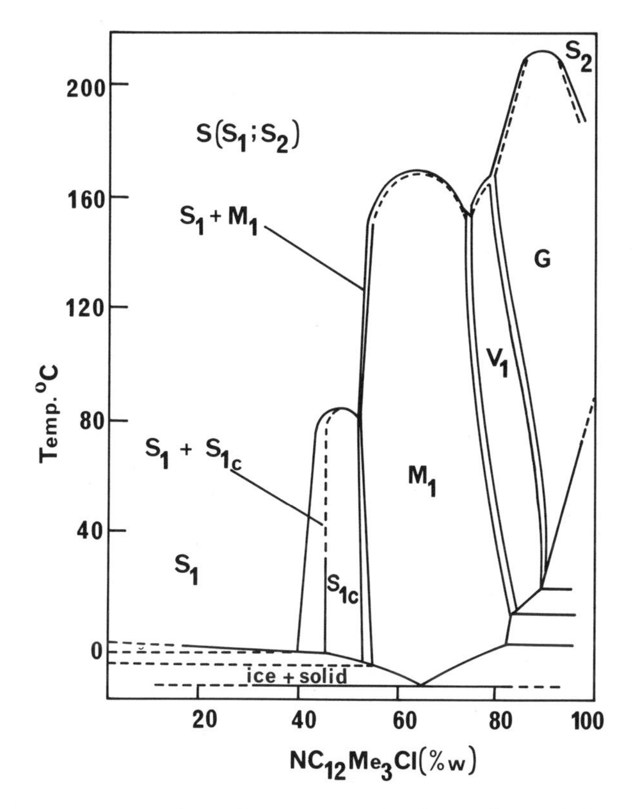

Figure 7. Phase diagram for the dodecyltrimethylammonium chloride ($NC_{12}Me_3Cl$)/water system (17)

ment of Cl^- by Br^- again results in displacement of the intermicellar equilibrium in Figure 1 to the right. Thus, Luzzati and Reiss-Husson (*18*) found by x-ray diffraction methods that in dodecyltrimethylammonium chloride solutions at 27°C, S_1 globular micelles persisted up to a concentration of about 40 wt %, while with the corresponding bromide the transformation to fibrous M_1 micelles occurred at a concentration of only 5%—*i.e.*, at an intermicellar separation distance too great for the formation of the S_{1c} lattice.

Formation of the V_1 and V_2 Mesophases and of the Non-Amphiphilic Cubic Mesophase "Smectic D". In Figures 1 and 5, the formation of the amphiphilic cubic mesophases V_1 and V_2 is attributed to transitional globular micellar forms which arise intermediate between the indefinitely extended fibrous (M) and lamellar (G) forms which constitute the middle and neat mesophases, respectively. With a few non-amphiphilic mesogens (*19, 20, 21, 22*) a cubic mesophase "smectic D" is found intermediate in the thermal succession of mesophases between smectic A

(Figure 2b) and smectic C (Figure 2c). It is now suggested that this may be caused by the formation of a rotational cubic lattice by globular transitional groupings of essentially parallel molecules produced as transitional intermediates, somewhat analogously to the transitional globular micelles of the V_1 and V_2 mesophases, between the indefinitely extended sheets of smectic A and smectic C types, respectively (Figure 2).

Literature Cited

1. Winsor, P. A., *Chem. Rev.* (1968) **68**, 1.
2. Winsor, P. A., *Mol. Cryst. Liquid Cryst.* (1971) **12**, 141.
3. "Liquid Crystals and Plastic Crystals," G. W. Gray, P. A. Winsor, Eds., Vol. 1, Chap. 5, Ellis Horwood Publishers, Chichester, England, 1974.
4. Luzzati, V., Mustacchi, H., Skoulios, A. E., *Discuss. Faraday Soc.* (1958) **25**, 43.
5. Mustacchi, H., Luzzati, V., Husson, F., *Acta Cryst.* (1960) **13**, 660.
6. Luzzati, V., Husson, F., *J. Cell. Biol.* (1962) **12**, 207.
7. Timmermanns, J., *Bull. Soc. Chim. Belg.* (1935) **44**, 17.
8. Timmermanns, J., *J. Phys. Chem. Solids* (1961) **18**, 1.
9. Gray, G. W., Winsor, P. A., *Mol. Cryst. Liquid Cryst.* (1974) **26**, 305.
10. Luzzati, V., Gulik-Krzywicki, T., Tardieu, A., *Nature* (1968) **218**, 1031.
11. Luzzati, V., Tardieu, A., Gulik-Krzywicki, T., Rivas, E., Reiss-Husson, F., *Nature* (1968) **220**, 485.
12. Tardieu, A., Luzzati, V., *Biochim. Biophys. Acta* (1972) **219**, 11.
13. Lawson, K. D., Mabis, A. J., Flautt, J., *J. Phys. Chem.* (1968) **72**, 2058.
14. Luzzati, V., Reiss-Husson, F., *Nature* (1966) **210**, 1351.
15. Winsor, P. A., "Solvent Properties of Amphiphilic Compounds," Butterworths, London, 1954.
16. Ekwall, P., Mandell, L., Fontell, K., *J. Colloid Interface Sci.* (1970) **33**, 215.
17. Balmbra, R. R., Clunie, J. S., Goodman, J. F., *Nature* (1969) **222**, 1159.
18. Reiss-Husson, F., Luzzati, V., *J. Colloid Interface Sci.* (1966) **21**, 534.
19. Gray, G. W., Jones, B., Marson, F., *J. Chem. Soc.* (1957) 393.
20. Demus, D., Kunicke, J., Neelson, J., Sackmann, H., *Z. Naturforsch.* (1968) **23a**, 84.
21. Diele, S., Brand, P., Sackmann, H., *Mol. Cryst. Liquid Cryst.* (1972) **17**, 163.
22. Gray, G. W., *Symp. Faraday Soc.* (1971) **5**, 94.
23. Aston, J. G., "Physics and Chemistry of the Organic Solid State," p. 543, Interscience Publishers, London, 1963.
24. Dunning, W. J., *J. Phys. Chem. Solids* (1961) **18**, 21.

RECEIVED November 19, 1974.

2

Lyotropic Mesomorphism

Phase Equilibria and Relation to Micellar Systems

INGVAR DANIELSSON

Department of Physical Chemistry, Åbo Akademi, Porthansgatan 3-5, SF-20500 Åbo (Turku) 50, Finland

A survey is given of lyotropic mesophase structures characterized by the special effect of the solvent in aqueous lipid systems. They are compared with micelle formation. The equilibria between homogeneous micellar solutions and mesophases follow Gibb's phase law. The solutions have no long range order while the structural parameters of the mesophases are determined by the compositions. The short range order, which reflects the internal structure, is the same for different aggregates. The mesophases have fixed transition temperatures. Polar groups are completely hydrated, but the mesophases also bind "intercalated liquid." Association in an aqueous environment is caused by hydrophobic interactions. Possibilities of analyzing molecular structure by NMR and Raman spectrometry are discussed. Attention is also drawn to the thermodynamic criteria influencing the formation of mesophases: of these only the activity of water is known so far.

The term lyotropic mesomorphism is used to describe the formation of thermodynamically stable liquid crystalline systems through the penetration of a solvent between the molecules of a crystal lattice. In contrast to the thermotropic mesomorphism shown by many pure substances, lyotropic mesomorphism always requires the participation of a solvent. Lyotropically mesomorphous systems, however, are usually as sensitive to changes in temperature as thermotropic systems. So far, lyotropic mesomorphism has been observed almost exclusively in lipid systems containing water. Lipids that show lyotropic mesomorphism frequently

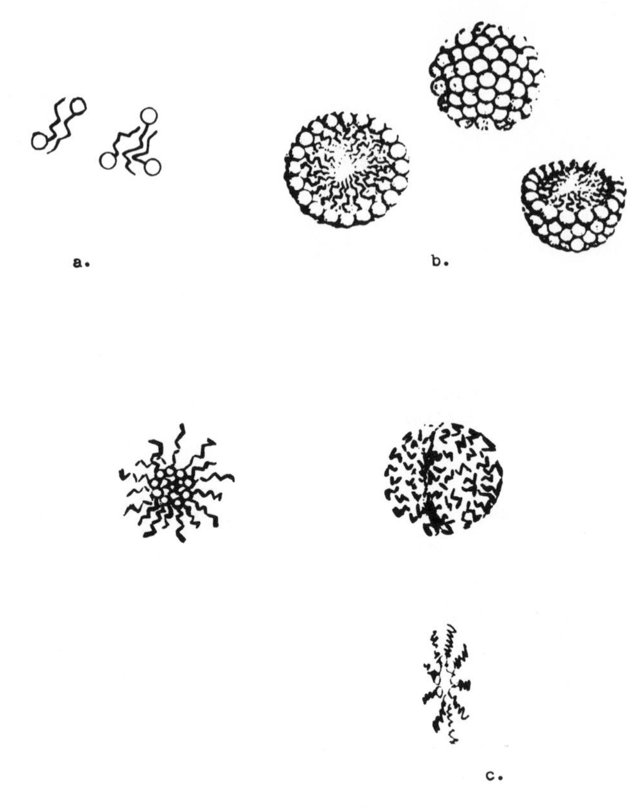

Figure 1. Structures of aggregates in solutions of association colloids

(a) Pre-micellar aggregates in water solution
(b) Ordinary micelles (Hartley micelles) in water solution
(c) "Inverted" micelles in oil solution

form thermotropic mesophases without any additions at high temperatures. Recent research concerning the lyotropic mesomorphism and its relation to micellar system covers a wide field, and only essential features can be given here.

The structure of the lyotropically mesomorphous lattice is made up of multimolecular units called mesoaggregates. These are surrounded by an intervening liquid. Lyotropic mesomorphism is therefore closely related to the tendency of lipids to accumulate at interfaces. The surface activity is a consequence of the same dualistic polar/non-polar molecular structure that causes the formation of micelles in solutions of association colloids (*1, 2, 3, 4, 5, 6*).

Micelles are large polymolecular aggregates in solutions. They are thermodynamically stable because of intermolecular interactions. Some types of micellar aggregates are illustrated in Figure 1. Small aggregates with few anions have been found in solutions of short-chain salts (Figure 1a). Normal Hartley micelles (Figure 1b) predominate in surfactant solutions above the critical micelle concentration (CMC). The lipophilic groups accumulate in the liquid-like inner part of the aggregates. The hydrophilic groups are directed towards water. "Inverted" micelles (Figure 1c) in a hydrocarbon environment have their polar groups piled up in the inner part of the micelles. These inverted micelles may also bind water. Small "inverted" aggregates, analogous to the premicelles formed in aqueous solutions, are frequently observed in organic solvents—for example, in extraction systems used in chemical technology.

When micellar aggregates are formed in solutions and their aggregation numbers are not very large, they are randomly dispersed, owing to thermal motion. Weak indications of anisotropy are found at very high concentrations only.

Structure of the Mesophases

X-ray investigations by Luzzati and Skoulios (*5, 7*) showed that the aggregates in lyotropic mesophases are structurally very similar to micelles (Figure 2). In contrast to these, however, the mesoaggregates, are usually of infinite dimensions in one or two directions. They form ordered lattices, which cause the characteristic anisotropy. The macroscopic flow properties of the mesophases depend on the fairly free translational mobility of the aggregates in one or two directions (Figure 2). The amphiphilic molecules are anchored through their polar groups on the borders between the mesoaggregates (type 2 E or 2 D) or in the regions in the mesoaggregates that are rich in water (type 2 F). The type 2 D lamellar phase consists of double layers of amphiphilic molecules—*i.e.*, their structure is analogous to that of bimolecular lipid membranes. Between these layers there is water; the hydrocarbon moieties of the molecules form liquid-like lipophilic inner layers that may solubilize oil.

The rod-shaped aggregates, 2 E, also have inner, lipophilic and outer, polar regions with the polar groups directed outwards. Usually the lattices are hexagonally arranged, but more complicated configurations have been encountered.

The aqueous region between the aggregates 2 E and 2 D is similar to an electrolyte solution. The electrical conductivity is good because of the fairly high mobility of the counterions, and the vapor pressure of the water is high since almost all long chain ions are bound (*4*).

There are no continuous water regions in mesophases of type 2 F. Their structure resembles that of inverted micelles. Their conductivity is low, and their properties are lipophilic.

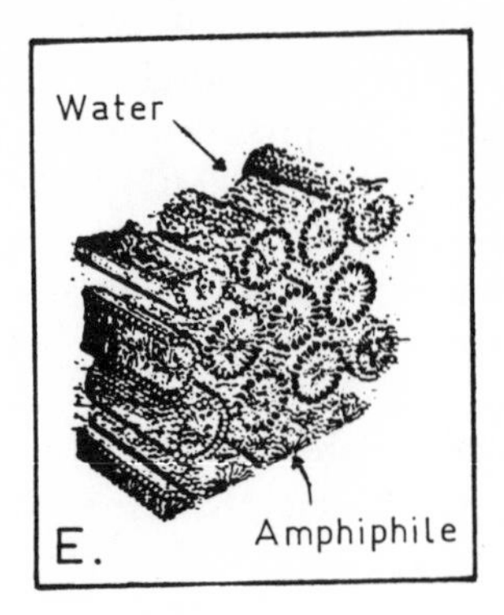

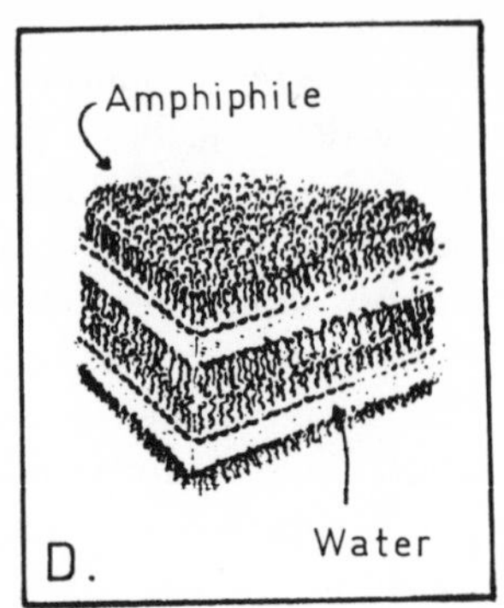

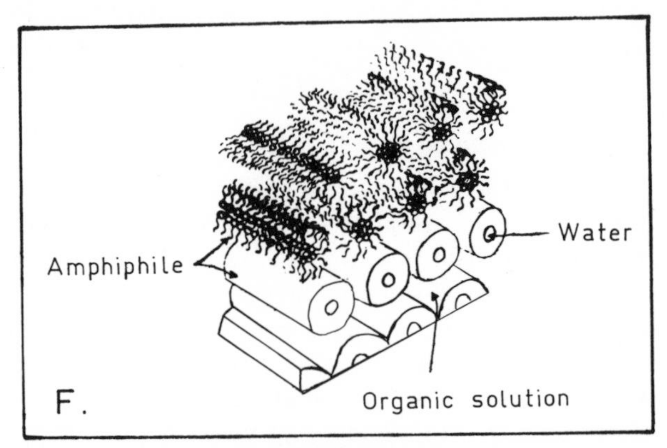

Figure 2. Some structures in liquid crystalline systems of association colloids

(E) Hexagonally packed hydrophilic rods with water-rich regions between the mesoaggregates (neat soap)
(D) Lamellar structure (middle soap)
(F) Hexagonally packed lipophilic rods with hydrophilic inner regions and oil-rich regions between the mesoaggregates.

The aggregates discussed above are all anisodimensional, which is the reason for the anisotropic character of the mesophases. In some systems it has been possible to prove the existence of isotropic highly viscous phases of similar structure but which clearly consist of almost isodimensional aggregates. The exact structure of these phases is still the subject of discussion, as is also the case with the complex mesophases. The relation between the isotropic phases and globular proteins and plastic crystals of non-amphiphilic substances has been discussed by Gray and Winsor (5).

The nomenclature for the different mesophases varies; translation lists for the terms used by different authors have been published by

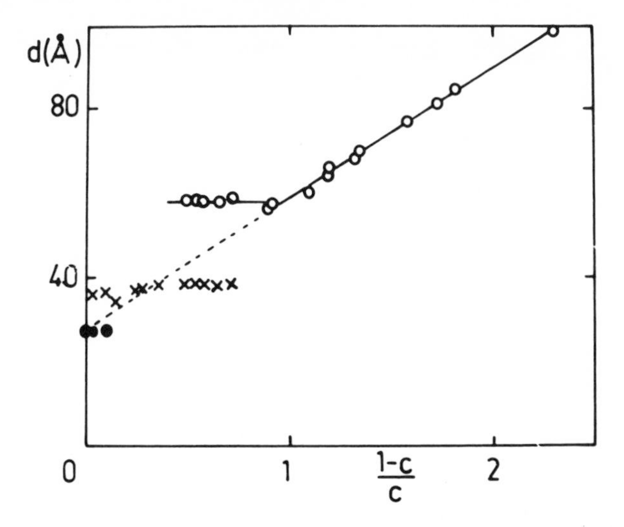

Acta Crystallographie

Figure 3. X-ray long spacings (A) of systems of cetyltrimethylammonium bromide and water as a function of the ratio gram water/gram association colloid (8)
○○○○ *lamellar phase*
×××× *hexagonal phase*
●●●● *crystals*

Ekwall and co-workers (*1, 2, 3*), as well as by others.

The macroscopic structure of lyotropic mesophases is reminiscent of emulsions that are stabilized by surfactants. Most often they are opalescent and gel-like. It is sometimes difficult to ascertain the difference between micellar and liquid crystalline systems on the one hand, and heterogeneous dispersions on the other. Heterogeneous, non-crystalline systems do not show x-ray diffraction as do homogeneous micellar and mesomorphous phases. The interplanear spacings corresponding to the aqueous and hydrocarbon regions in the mesophases vary with the volume fraction of water, as predicted by the structural models (Figure 3) (*8*). The mesophases have well-defined transition temperatures and occur in well-defined composition regions (Figures 3 and 4) (*8, 9*).

The state of the hydrocarbon chains in mesophases and micelles is reflected in the Krafft phenomena. In aqueous solutions of surfactants the Krafft point is defined as the temperature at which the solubility reaches the critical micelle concentration; when the temperature is increased further, the solubility rises rapidly since the monomers form micelles (Figure 5) (*10*). Lipids that do not form micelles frequently start to swell by the uptake of water at a well-defined temperature; they are transformed into a mesomorphous state (Figure 6) (*11*). The relation between these two Krafft phenomena is explained to some extent by the

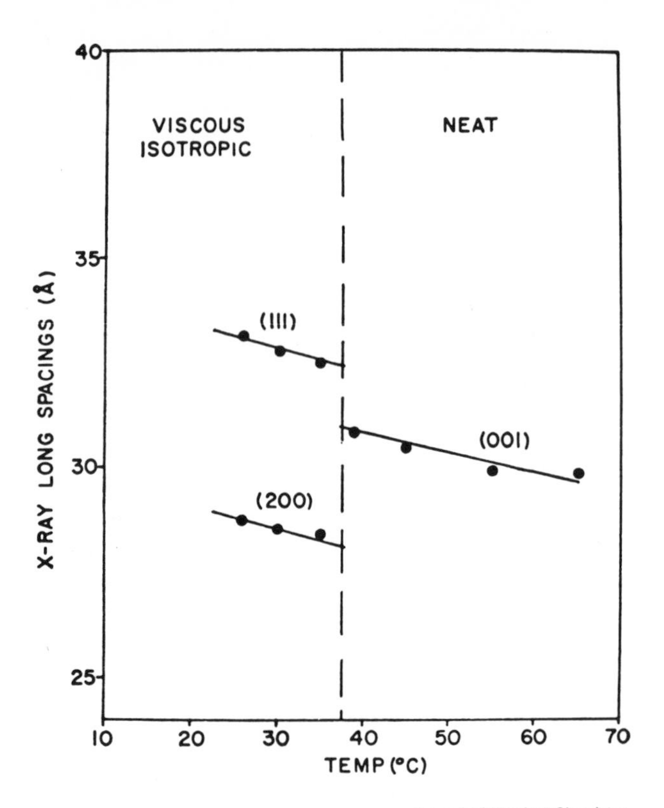

Journal of Physical Chemistry

Figure 4. X-ray long spacings of a 69.6% dimethyldodecylamine oxide in deuterium oxide solution as a function of temperature (9)

fact that the aqueous layer around the polar groups of the surfactant facilitates the transition of the lipophilic palisade layer to the liquid-like state that is characteristic of the mesophases (*1, 2, 3*). A weakened hydrocarbon–hydrocarbon interaction caused by bulky hydrocarbon groups or double bonds has a similar effect (*11*).

Phase Diagrams

The phase equilibria and diagrammatic structures in a two-component system are illustrated in Figure 7 (*12*). The phase diagrams have often been determined by inspection with the naked eye. More exact methods include x-ray and NMR analysis, density measurements, separation by high-speed centrifugation or differential thermal analysis.

Ternary systems of surfactant, slightly polar additives and water are particularly interesting. A typical example is the classical model system

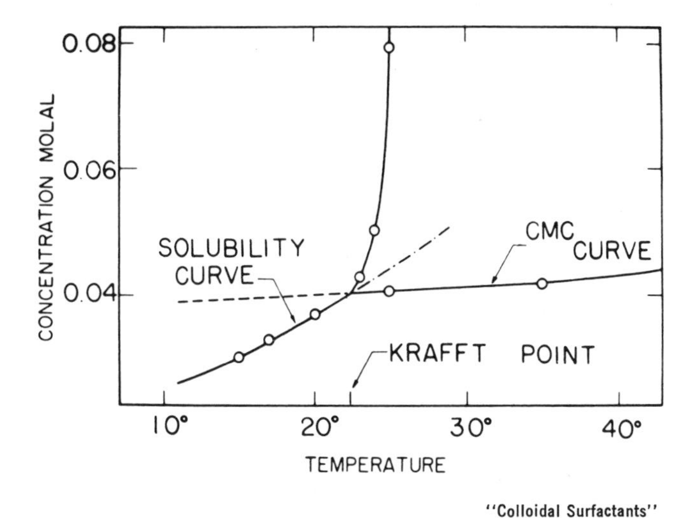

"Colloidal Surfactants"

Figure 5. Phase diagram close to the Krafft point (10)

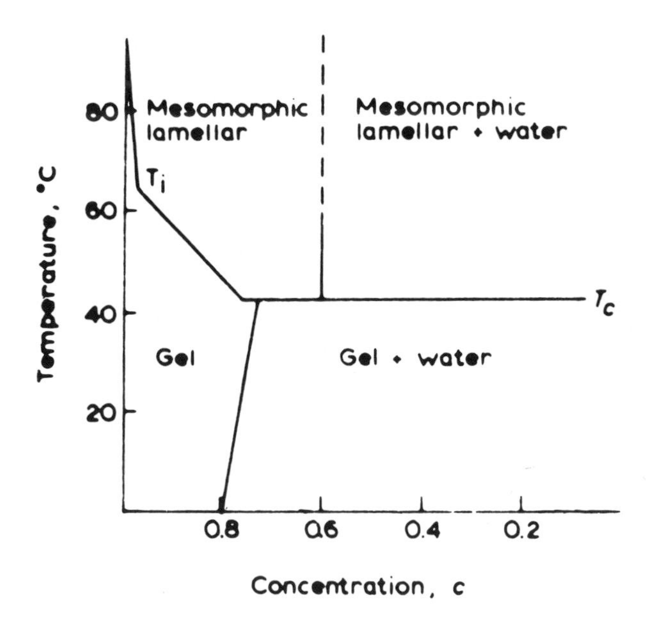

"Form and Function of Phospholipids"

Figure 6. Phase diagram of the 1,2-dipalmitoyl-glycero-3-sn phosphorylcholin–water system (11)

of sodium octanoate–decanol–water at 20°C which has been studied by Ekwall and co-workers (Figure 8) (*13*). Some diagrammatic structures are shown; all multiphase regions have been left out. In the system, one finds normal Hartley micelles (L_1), inverted micelles (L_2), lamellar

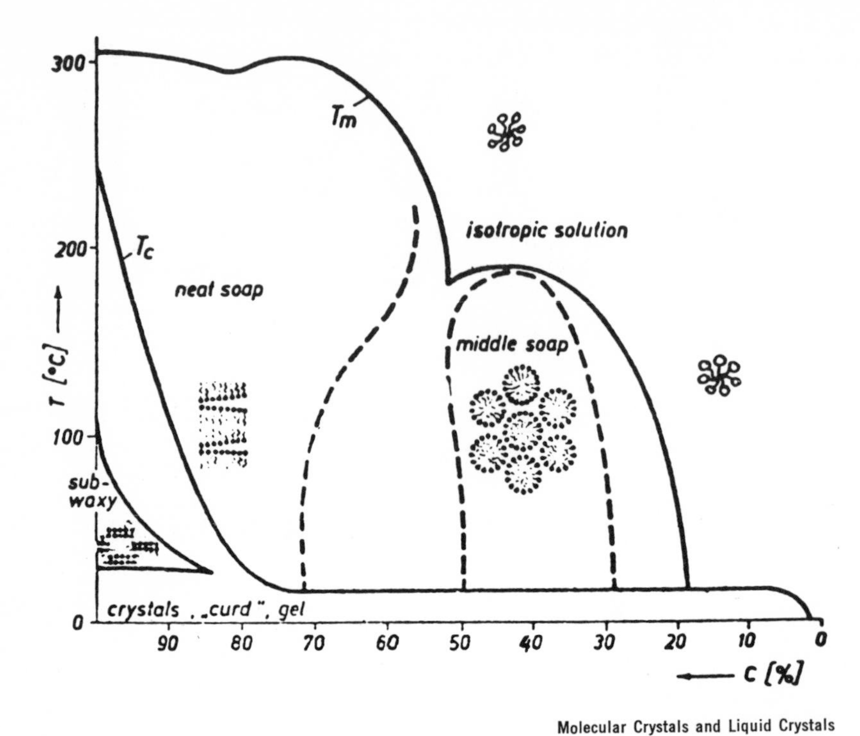

Molecular Crystals and Liquid Crystals

Figure 7. Schematic of a soap phase diagram and structure of some phases. The boundaries for the main phases of the system are 30–60% for the hexagonal phase and 71–82% for the lamellar phase at 20°C (12).

structures (D), mesophases consisting of polar rods in an aqueous continuum (E), and non-polar rods in a hydrocarbon continuum (F). The structures of phases B and C are not definitely settled.

The same phase equilibria are shown in Figure 9, but here the two-phase and three-phase regions have been introduced. There are eight three-phase triangles representing the equilibria L_1-L_2-D, L_1-B-D, L_1-C-D, L_1-E-D, L_2-D-F, D-E-G, and L_2-F-G. In the region of the triangular diagram close to the crystalline sodium octanoate the very small amounts of solution and mesophases make it difficult to separate the equilibrium phases. In region G the external appearance of the soap is crystalline or curd-like; the soap includes considerable amounts of decanol and water. The heterogeneous systems between L_1, B, and C do not separate well on centrifugation. The compositions vary with the strength and duration of the centrifugal field. In regions B and C, the forces between the mesoaggregates are possibly weak enough to be influenced by centrifugation. These phase equilibria are also extremely sensitive to temperature.

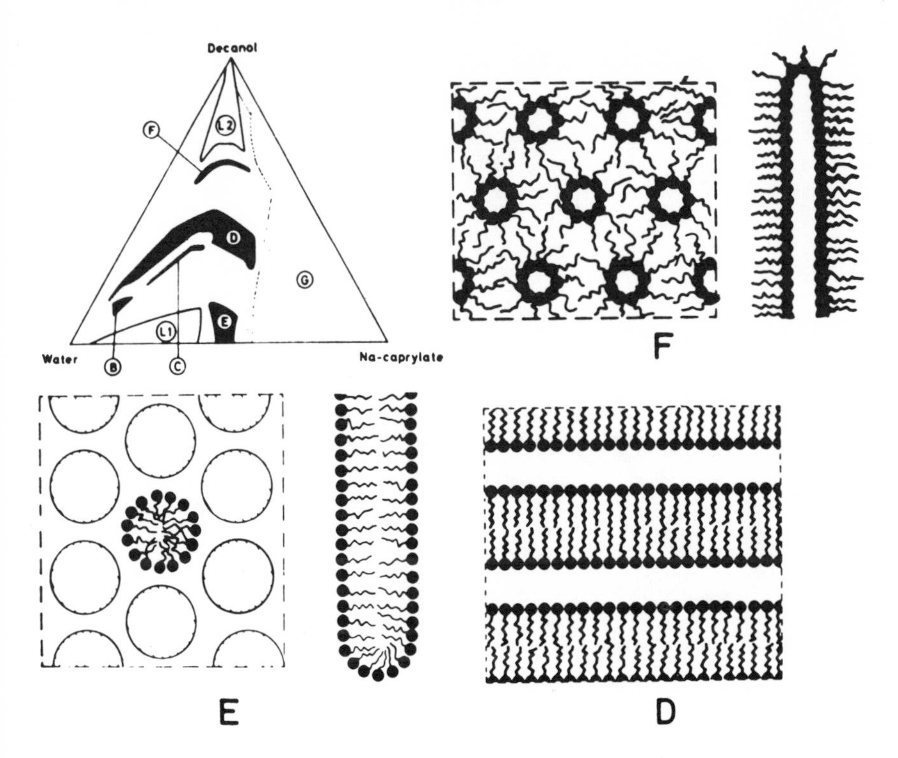

"MTP International Review of Science"

Figure 8. Schematic of mesomorphous structures in phase E, D, and F in the three-component system: sodium octanoate–decanol–water (1)

The exact extension of the different phase regions in the diagram is less interesting than the fact that the lyotropic lipid systems always follow the phase rule from a macroscopic point of view. In the ternary diagram (Figure 9) there are never more than three phases present at any one time. Each three-phase triangle is surrounded by two-phase regions, and the corners of the three-phase regions end in one-phase regions. This shows that the mesophases are really homogeneous equilibrium systems; they are not, for example, highly dispersed emulsions.

Since the mesophases may contain more than 90% water, the presence of a fourth component, even as a small impurity, may distort the whole phase diagram. The requirements of the phase rule were often neglected in earlier studies, but even in recent papers contradictory diagrams can be found.

Because of the microscopic and macroscopic validity of the phase rule the micelles in the isotropic solutions L_1 and L_2 must be regarded as polymolecular complexes in solution and not as a separate micellar phase.

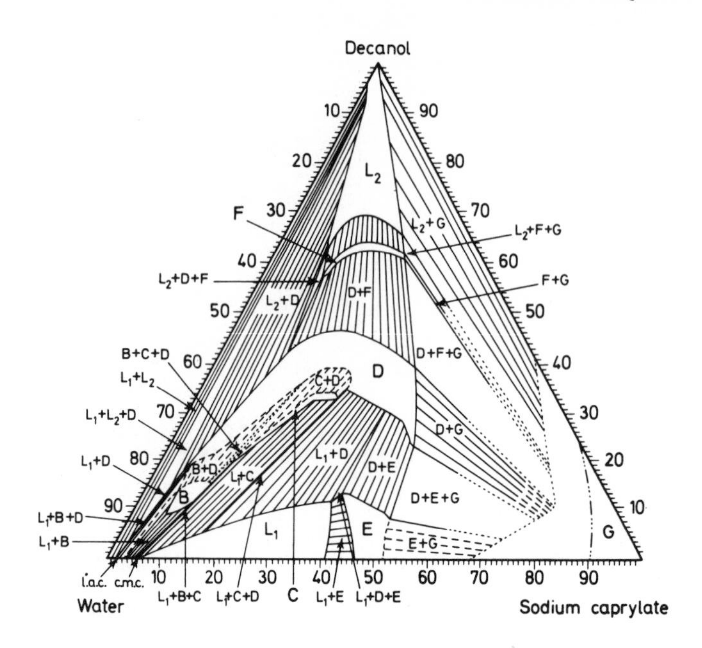

"MTP International Review of Science"

Figure 9. Phase diagram for the three-component system: sodium octanoate–n-decanol–water at 20°C (1). Concentrations expressed as weight percent.

L_1: Region with homogeneous, isotropic aqueous solution.

L_2: Region with homogeneous, isotropic decanolic solution.

B, C, D, E, F: Regions with homogeneous mesomorphous phases.

E: Region with homogeneous mesophase with two-dimensional, hexagonal structure; amphiphilic rods; "normal" structure.

F: Region with homogeneous mesophase with two-dimensional hexagonal structure; water rods; "reversed" structure.

D: Region with homogeneous mesophase with lamellar structure; one-dimensional swelling.

Thermodynamics

To characterize the mesophases thermodynamically, it is desirable to have detailed knowledge of the regions in which phases B and C exist in the model system sodium octanoate–decanol–water. The heats at which the mesophases are formed are somewhat easier to obtain than the chemical potentials. Table I gives some examples of the few calorimetric values available (*14, 15*).

The heat of solubilization, heat of micelle formation, and heats of formation of phases D and E are small in comparison with ordinary

Table I. Apparent Heats of Aggregation in Micellar Solutions and Mesomorphic Phases in the System: Sodium Octanoate–*n*-Pentanol–Water (*14, 15*)

Aqueous Solution	*Aggregation*	ΔH
	Micelle formation	− 7.6 kJ/mol
30% $NaOOCC_7H_{15}$	Transfer of *n*-pentanol from water to micelles	+ 4.5 kJ/mol
	Formation of phase E from saturated solution	Δl_{NaC_8} = 23 J/mol Al_{H_2O} = −300 J/mol

chemical reactions. The entropy of the change in Gibbs energy on aggregation is therefore of the same order as the enthalpy. The formation of micelles and mesoaggregates from their components in standard states is therefore governed by similar factors.

(a) The hydrocarbon chains have more torsional freedom in the aggregates than in the rigid molecular configuration stabilized by the surrounding water (*16*).

(b) The entropy increases as the structured water around the hydrocarbon chains returns to "normal" behavior. This is sometimes expressed, rather inadequately, as an "expulsion of hydrocarbon chains from water" (*17*).

The binding of counterions and van der Waals' attraction make further contributions to the association. The heats of formation determined so far, however, are insufficient. More definite knowledge about the nature of the binding has been obtained by spectroscopic investigation.

So far it has not been possible to measure the chemical potentials of the components in the mesophases. This measurement is possible, however, in solutions which are in equilibrium with the mesophases. If pure water is taken as the standard state, the activity of water in equilibrium with the D and E phases in the system NaC_8–decanol–water is more than 0.8 (*4*). From these activities in micellar solutions, the activity of the fatty acid salt has sometimes been calculated. The salt is incorrectly treated as a completely dissociated electrolyte. The activity of the fatty acid in solutions of short chain carboxylates has also been determined by gas chromatography; from these determinations the carboxylate anion activity can be determined (*18*). Low CMC values for the carboxylate are obtained (*15*). The same method has shown that the activity of solubilized pentanol in octanoate solutions is still very low when the solution is in equilibrium with phase D (Figure 10) (*15*).

The kinetic micellar units include about half of the counterions; the mesoaggregates bind counterions even more firmly. It can be estimated from activity measurements, however, that a considerable proportion of

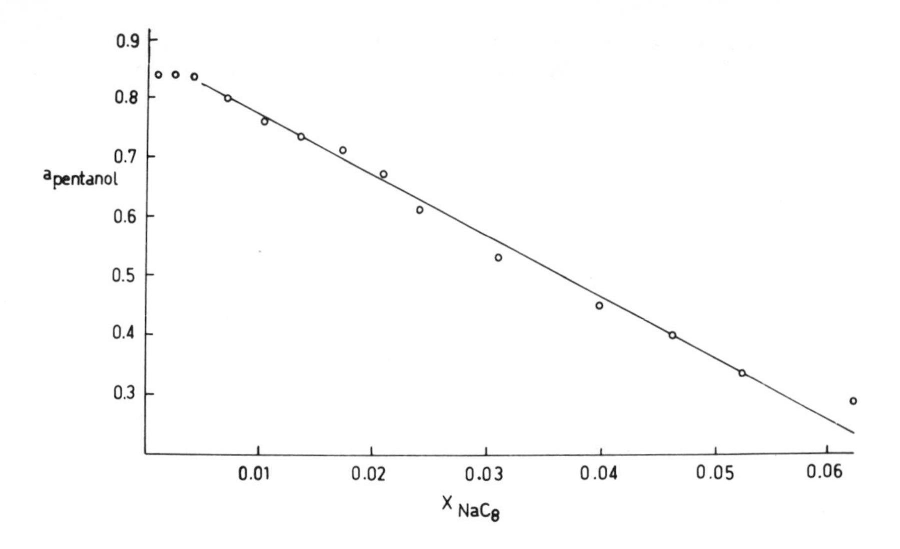

15th Scandanavian Chemistry Meeting

Figure 10. The activity of n-*pentanol in micellar solutions of sodium octanoate with solubilized pentanol and in equilibrium with the lamellar mesophase D. The abscissa indicates the mole fraction of sodium octanoate in the system* (15).

the sodium ions remain free; a certain mobility and conductivity are retained. This dissociation gives rise to a Donnan potential, which, according to Ekwall, explains the extreme water-binding capacity of some mesophases (*4*).

The chemical potentials measured so far do not allow the formulation of thermodynamic criteria for the formation of lyotropic mesophases. Some qualitative remarks, however, can be made. Of particular interest are Ekwall's studies of the relations between the water binding of the mesophases, their ionization, x-ray parameters, and vapor pressures (*4*). For common soaps at room temperature mesophases can be observed only in the presence of amounts of water that hydrate the ionic and polar groups. Hydration is therefore characteristic of aqueous lyotropic mesophases as well as micellar systems (*1, 2, 3*). The binding of counterions to the micelles and to the mesoaggregates seems to be of a similar electrostatic nature. The addition of NaCl greatly affects the lamellar phase D and, to a lesser extent, phase E; in these phases the counterions are more strongly bound than by micelles in the solution (*1, 2, 3*).

In the ternary system NaC_8–decanol–water the influence of the polar/apolar solubilizate on the formation of micelles and mesoaggregates can be seen clearly (Figure 9). Quite often the following rules of thumb for the influence of the solubilizate can be used:

(a) Non-polar solubilizates are built into the lipophilic moiety of the aggregates with little influence on the aggregation numbers. The hydrophilic surfaces of the aggregates and the aqueous solubility of the surfactants are not changed; excess solubilizate separates as a pure substance.

(b) Polar/apolar additives are built into the palisade layers and strongly influence the charge density, the counterion binding, and the interaction with water of the micelles and the mesoaggregates. From aqueous solutions, excess solubilizate separates as a mesophase with a high water content. Non-electrolyte polar additives in particular induce the formation of lamellar mesophases, owing to the decrease in the surface charge density of the aggregates. The formation of rod-like mesoaggregates of type E is induced when the charge density is high.

(c) Surfactants in weakly polar solvents form inverted micelles with the water solubilized in strongly polar inner environments. From these solutions there is also frequently a separation of mesophases that consists of lamellar or non-polar rodlike aggregates. In these aggregates the water is bound as hydration water and cannot be compared with an aqueous solution. Examples of these cases are given by the two-phase equilibria F-L_2 and F-D in Figure 9.

In two-component systems of association of colloid and water the sequence of phases, as the water content decreases, is: micellar solution $\rightarrow$ hexagonally packed polar rods $\rightarrow$ complex phases with rod-shaped aggregates $\rightarrow$ lamellar mesophase D $\rightarrow$ crystalline surfactant. Some of these steps may be absent, depending, for example, on the temperature.

Reactions between Various Association Structures

We now turn to the question of whether there is any successive relationship involved in lower complexes, micelles, and the different mesoaggregates. Such a succession can formally be attributed to the tendency of the aggregate surfaces to be convex when they are in contact with water or a hydrocarbon in accordance with the R theory developed by Winsor (*5, 6*). The association processes as such are, however, very fast. Present experimental techniques show that association can be considered as a series of second-order reactions, the surfactant ions being successively caught at about similar rates until the optimal size of the aggregates has been reached (*6*). The dissociation of the aggregates is also very fast and is a first-order reaction. We have only investigated the stationary equilibria and, so far, no one has been able to decide whether the mesoaggregates are formed by the restructuring of smaller aggregates or micelles. We can only conclude that lyotropic mesomorphism is almost always shown by micelle-forming surfactants; however, lyotropic mesophases can frequently be formed without any association in the aqueous solution.

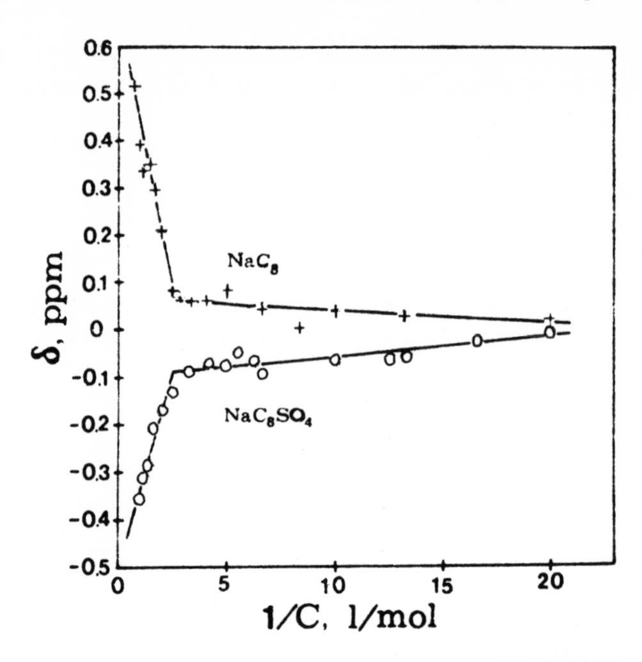

"Liquid Crystals and Ordered Fluids"

Figure 11. [23]Na chemical shifts (in ppm) for isotropic solutions of sodium octanoate (×) *and sodium octylsulfate* (○) *as a function of the inverse soap concentration. A positive δ denotes a shift to lower field* (23).

The most important new results concerning the structure of micelles and mesophases in recent years have been achieved using NMR methods. These have been particularly successful when studying the changes undergone by groups of hydrocarbons when transferred from an aqueous environment to micelles and mesoaggregates and in the binding of water and counterions to the aggregates (*19–25*). As examples of studies of this kind Figure 11 describes chemical changes of ^{23}Na in isotropic solutions of sodium octanoate and sodium octylsulfate both above and below the critical micelle formation concentration (*21, 22, 23*).

Laser-Raman spectroscopy is a new method with considerable potential for providing an explanation of how the surroundings inside the aggregates influence the crystalline state of the hydrocarbon chains and other groups (*25, 26*). It seems probable, however, that an important area of research on the phase equilibria proper would concentrate on attempts to throw light on the exact thremodynamic criteria for the association processes. Ekwall's studies of the water activities of mesophases in the system water–decanol–sodium caprylate are an example of such research (*4*). However, thermodynamic treatment of the association processes presupposes measurements of the activities of several different components in the systems. When supplemented by measurements of the heats of formation, such studies should provide information about the entropy effects in the systems and consequently help to explain the fundamental causes of the association phenomenon. Such studies would also be very important in helping to interpret spectroscopic results.

Literature Cited

1. Ekwall, P., Danielsson, I., Stenius, P., "MTP International Review of Science," Physical Chemistry Series I, Vol. 7, p. 97, A. D. Buckingham and M. Kerker, Eds., Butterworths, London, 1972.
2. Ekwall, P., Stenius, P., "MTP International Review of Science," Physical Chemistry Series 2, Vol. 7, in press.
3. Ekwall, P., *Adv. Liq. Cryst.* (1975) **1,** 1.
4. Ekwall, P., "Liquid Crystals and Ordered Fluids," J. F. Johnson and R. S. Porter, Eds., Vol. 2, p. 177, Plenum, New York, 1974.
5. Gray, G. W., Winsor, P. A., *Mol. Cryst. Liq. Cryst.* (1974) **26,** 305.
6. Winsor, P. A., *Mol. Cryst. Liq. Cryst.* (1971) **12,** 141.
7. Luzzati, V., in "Biological Membranes," D. Chapman, Ed., p. 71, Academic, London, 1968.
8. Vincent, J. M., Skoulios, A., *Acta Cryst.* (1966) **20,** 444.
9. Lawson, K. D., Mabis, A. J., Flautt, T. J., *J. Phys. Chem.* (1968) **72,** 2058.
10. Shinoda, Kozo, Nakagawa, Toshio, Tamamuchi, B.-I., Isemura, Toshizo, "Colloidal Surfactants," p. 7, Academic, London, 1963.
11. Chapman, D., Williams, R. M., "Form and Function of Phospholipids," B.B.A. Library 3, G. B. Ansell, R. M. C. Dawson, and J. N. Hawthorne, Eds., p. 126, Elsevier, New York, 1973.
12. Eins, S., *Mol. Cryst. Liq. Cryst.* (1970) **11,** 119.
13. Ekwall, P., Mandell, L., Fontell, K., *Mol. Cryst. Liq. Cryst.* (1969) **8,** 157.
14. Danielsson, I., *Proc. Scandinavian Symp. Surface Chem., 5th, Abo, 1973.*
15. Rosenholm, B., *Proc. Scandinavian Chem. Meetg., 15th, 1974,* p. 64.
16. Aranow, R. H., Whitten, L., *J. Phys. Chem.* (1960) **64,** 1643.
17. Poland, D. C., Sheraga, H. A., *J. Phys. Chem.* (1965) **69,** 2431.
18. Backlund, S., Danielsson, I., *Proc. Intern. Congr. Surface Active Substances, 6th, Zürich, 1972,* p. 1013.
19. Rassing, J., Wyn-Jones, E., *Chem. Phys. Lett.* (1973) **21,** 93.
20. Johansson, A., Lindman, B., in "Liquid Crystals and Plastic Crystals," G. W. Gray and P. A. Winsor, Eds., Ellis Horwood, Chichester, 1974.
21. Tiddy, G. J. T., *J. Chem. Soc. F1* (1972) **68,** 369.
22. *Ibid.,* p. 653.
23. Gustafsson, H., Lindblom, G., Lindman, B., Persson, N.-O., Wennerström, H., "Liquid Crystals and Ordered Fluids," J. F. Johnson and R. S. Porter, Eds., Vol. 2, p. 161, Plenum, New York, 1974.
24. Wennerström, H., Lindblom, G., Lindman, B., Arvidsson, G., *Chem. Phys. Letters* (1974) **12,** 4, 261.
25. Larsson, K., Rand, R. P., *Biochem. Biophys. Acta* (1973) **326,** 245.
26. Parsegian, V. A., *Trans. Faraday Soc.* (1966) **62,** 848.
27. Larsson, K., *Chem. Phys. Lipids* (1972) **9,** 181.

RECEIVED November 19, 1974.

3

Micellar and Lyotropic Liquid Crystalline Phases Containing Nonionic Active Substances

STIG FRIBERG and LISBETH RYDHAG

The Swedish Institute for Surface Chemistry, Drottning Kristinas väg 45, S-114 28 Stockholm, Sweden

TADASHI DOI

Kao Soap Ltd., Tokyo, 103 Japan

Phase diagrams of water, hydrocarbon, and nonionic surfactants (polyoxyethylene alkyl ethers) are presented, and their general features are related to the PIT value or HLB temperature. The pronounced solubilization changes in the isotropic liquid phases which have been observed in the HLB temperature range were limited to the association of the surfactant into micelles. The solubility of water in a liquid surfactant and the regions of liquid crystals obtained from water–surfactant interaction varied only slightly in the HLB temperature range.

The phase equilibria of ionic surfactants combined with water and an amphiphilic substance such as a long chain alcohol, carboxylic acid, or ester have been investigated in detail for a long time (*1*, *2*). The nonionic surfactants have not attracted as much interest despite the fact that they are suitable models for illustrating the association conditions which are responsible for the structure and function of biomembranes; they also present interesting problems in their temperature dependent interaction with water and hydrocarbons.

This article discusses some micellar and liquid crystalline phases with nonionic substances, water, and hydrocarbons; and some factors are delineated for their association phenomena. Lipid phase behavior has an extremely important direct influence on certain biological phenomena (Chapter 10) and is treated in Chapter 4. The treatment here is limited to one class of compounds—the nonionic surfactants containing a polyoxyethylene chain as the hydrophilic part in addition to the hydrophobic hydrocarbon part.

These compounds differ from other surfactants in the pronounced sensitivity of their association structural organization to temperature. This characteristic feature was noted very early by Shinoda (*3*) with regard to their micellar association and solubilization. A corresponding sensitivity may also be observed in the strong dependence of the liquid crystalline regions in phase diagrams of the system: water, surfactant, and hydrocarbon (*4*).

Figure 1 demonstrates the drastic influence on the stability region of a lamellar liquid crystalline phase when an aromatic hydrocarbon is substituted by an aliphatic one. The lamellar phase formed by water and emulsifier is stable between 20 and 60 wt % water. Addition of an aromatic hydrocarbon (*p*-xylene) to the liquid crystalline phase increased the maximum amount of water from 45 to 85% (w/w) (Figure 1 left). Inclusion of an aliphatic hydrocarbon (*n*-hexadecane) gave the opposite result; the maximum water content in the liquid crystalline state was reduced (right). Some of the factors which govern the association behavior of these surfactants and cause effects such as the one above are treated below.

Water–Polyoxyethylene Chain Interaction

The ether bridges in the polyoxyethylene chain interact with the water molecules through weak hydrogen bonds (*5*). Quantum mechanical calculations have demonstrated the shallow energy minimum experi-

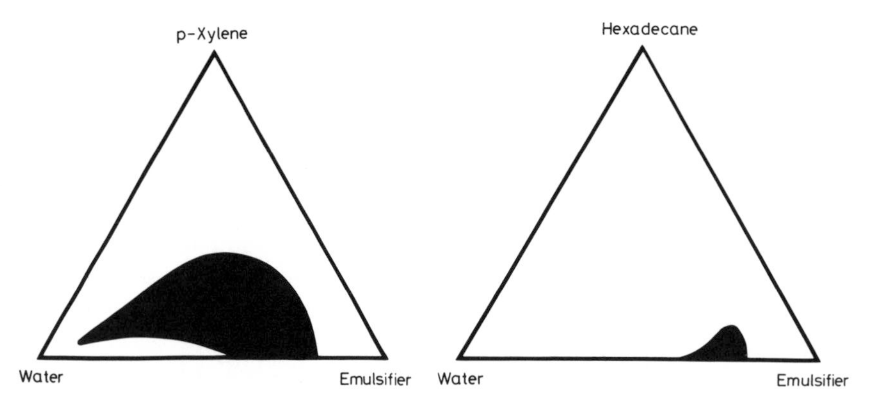

Figure 1. Difference in the phase region of the lamellar liquid crystal (black) when an aromatic hydrocarbon (left) is replaced by an aliphatic one (right); demonstrates the sensitivity of the lyotropic liquid crystalline structure to weak intermolecular forces. The emulsifier is a polyoxyethylene (9) nonyl phenol ether.

enced in this interaction. The recent study by Ray and Némethy (*6*) on micellization in the presence of ethylene glycol exposed the combined hydrophobic–hydrophilic interactions of the polyoxyethylene chain.

Direct interaction through the hydrogen bond does not completely explain the solubility of the polyoxyethylene chain in water; this condition is well illustrated by the fact that in contrast to the polyoxyethylene compounds neither the corresponding polyoxypropylene nor the polyoxymethylene substances showed a similar solubility in water (*7*). This occurred despite the fact that the polyoxymethylene chain is more hydrophilic as exemplified by calculations of the number for hydrophilic lipophilic balance (HLB) from group numbers (*8*). The oxypropylene group has a group number of −0.15 which implies a hydrophobic character. The corresponding number for the oxyethylene group is +0.33 which illustrates a hydrophilic tendency. The oxymethylene group number is considerably more positive, +0.8, suggesting strong hydrophilic behavior. Despite this positive number the polyoxymethylene chain does not confer water solubility on the compound.

The reason for this anomalous behavior is not clear, but it might involve the structural dimensions of the polyoxyethylene chain. It appears generally accepted (*9*) that the chain attains a helix formation in water with seven groups per helix revolution—*i.e.*, a 7_2 helix. The oxygen–oxygen distance in this chain is 0.277 nm, which is exactly the oxygen–oxygen distance in the tetrahedrically hydrogen bonded wurtzite structure of water. This fact combined with the strong temperature dependence of the solubility of nonionic surfactants in water (*vide infra*) suggests a possible relationship between a long range order portion of the water structure and the solubility of the polyoxyethylene chain.

Micellization and Solubilization

The micellization of surfactants has been described as a single kinetic equilibrium (*10*) or as a phase separation (*11*). A general statistical mechanical treatment (*12*) showed the similarities of the two approaches. Multiple kinetic equilibria (*13*) or the small system thermodynamics by Hill (*14*) have been frequently applied in the thermodynamics of micellization (*15, 16, 17*). Even the experimental determination of the factors governing the aggregation conditions of micellization in water is still a matter of considerable interest (*18, 19*) and dispute (*20*).

The influence on micellization caused by additives in the water has been investigated to a lesser extent (*7, 21, 22, 23*), and the mechanism of the influence, even for small nonelectrolyte molecules (*23*), has not yet been described unambiguously. Micellization and solubilization in solvents other than water have been investigated (*8, 24, 25, 26, 27, 28*).

These studies throw light on the initial aggregation phenomena, which results in micelles and may also (*17*) constitute the necessary basis for understanding the subsequent agglomeration to liquid crystals. Very little is known about the thermodynamic conditions for the latter associations. Instead we must rely on empirical data to illuminate the basic mechanisms which determine the association behavior of those substances in concentrated systems. Valuable information for understanding the drastic influence of weak intermolecular forces on the association structures (Figure 1) is obtained from the pronounced temperature dependence of the solubilization which was observed early by Shinoda (*29*). He and his collaborators (*30, 31, 32 33*) have since developed this subject.

Temperature Dependence and Solubilization

While the amount of solubilized hydrocarbon changes very little with temperature when ionic surfactants are used, the solubilization capacity of nonionics is drastically influenced by temperature. Figure 2 (*29*) shows the temperature dependence of the solubilization of *n*-heptane, cyclohexane, and ethylbenzene in a 1% (w/w) aqueous solution of polyoxyethylene (9.2) nonyl phenol ether. In general, solubilization increases in this manner close to the cloud point with increasing temperature. Similarly the solubilization of water in hydrocarbons increases when the temperature is reduced (*25, 26*). In a narrow temperature range the solubilization behavior consequently changes from oil in water (O/W) to water in oil (W/O) with increasing temperature. The temperature range in which this change is experienced was termed the phase inversion

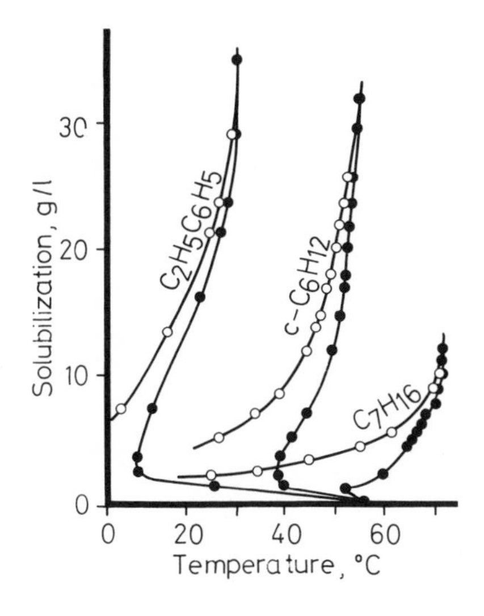

Figure 2. Solubilization of three hydrocarbons by polyoxyethylene (9.2) nonyl phenol ether in water as function of temperature (29)

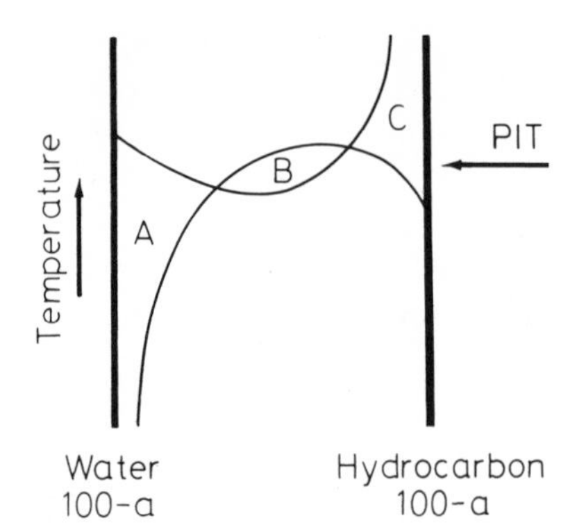

Figure 3. General features of micellar phases in the system: water, hydrocarbon, and a nonionic surfactant (wt %). A: O/W solubilization; B: surfactant phase; C: W/O solubilization.

temperature (PIT) by Shinoda, and he described its dependence on the cloud point and the molecular interactions (*30*).

In the PIT range Shinoda observed an isotropic liquid phase called the surfactant phase (*31*); the general features of the micellar solution regions are illustrated in Figure 3. The influence of several factors on the size of the various regions has been amply described (*32, 33*).

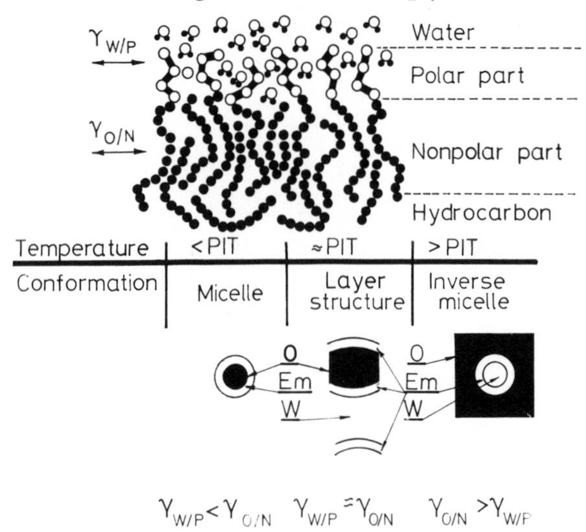

Figure 4. Formation of normal micelles (left), the surfactant phase (middle), and inverse micelles (right) may be referred to the relative size of the interfacial tensions against the oil ($\gamma_{O/N}$) and the water ($\gamma_{W/P}$) at a plane interface (32)

One interfacial tension (upper left) is considered located between water and the polar parts (unfilled circles) of the surfactant (upper right) and one (middle left) between the nonpolar part (filled circles) of the surfactant and the hydrocarbon (middle right). The different convexities of the O/W interface giving normal micelles, a surfactant phase or an inverse micelle are formally referred to different ratios of these interfacial tensions (bottom of figure) at a plane interface.

Shinoda rationalized the results in terms of a divided total interfacial tension: one between the polar parts of the surfactants and the water ($\gamma_{W/P}$), and one between the hydrocarbon and the nonpolar part of the surfactant ($\gamma_{O/N}$); similar reasonings have been used by Prince (*34, 35*) and Robbins (*36*) (Figure 4). Essentially the theory states that at low temperatures the interfacial tensions at a plane interface

$$\gamma_{W/P} < \gamma_{O/N}$$

while at higher temperatures the reverse is true. In a temperature range near the PIT

$$\gamma_{W/P} \simeq \gamma_{O/N}$$

favoring a plane interface. Shinoda concluded that a layer structure should be obtained in which the surfactant was continuous and the oil and water were solubilized next to the hydrophobic and hydrophilic parts (Figure 4).

The surfactant phase was observed as a separate solubility region in the three-component system: water, emulsifier, and hydrocarbon (*37*). The composition of the phase was extremely sensitive to temperature (*see* Figure 5). In general the solubility regions vary according to Figure 6. The PIT region is characterized by a minimum of the emulsifier concentration necessary to obtain an isotropic liquid solution of concurrently large amounts of water and hydrocarbon.

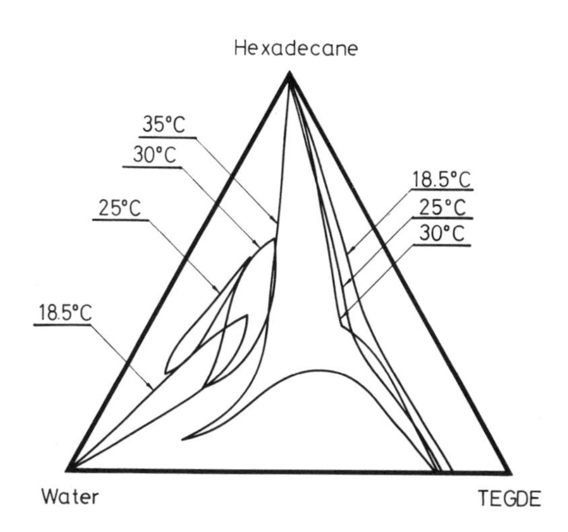

Figure 5. Phase regions for isotropic solutions in the system: water, hexadecane, and tetraoxyethylene dodecyl ether (TEGDE) (37)

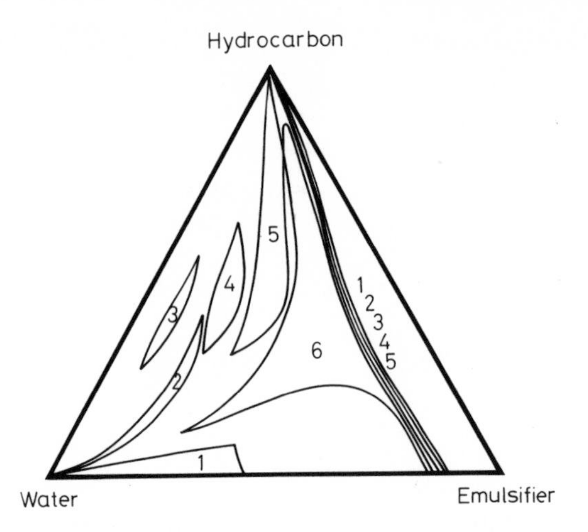

Figure 6. General features of the isotropic solution regions in a system: water, hydrocarbon, and nonionic emulsifier with an oxyethylene chain length of about 4. The numbers 1–6 show the regions at increasing temperatures.

If the liquid crystalline phase is included in the diagram, the general features are those in Figure 7 (*38*). At this temperature (the PIT or HLB temperature) increasing amounts of emulsifier first give rise to an isotropic liquid (S) in a small concentration range (A-B), followed by a phase transition to a lamellar liquid crystal (N) in the concentration range C-D.

In this context it is instructive to ruminate on the structure of the surfactant phase. A representative composition of the phase would be 10% emulsifier and equal amounts of water and hydrocarbon. The conclusions giving a layer structure (*31, 32, 33*) appear to be a reasonable basis for discussing the energy conditions implied in the structure. If an area per molecule of 10^{-18} m^2 is considered reasonable (*39*), the water and oil layers are approximately 1.2×10^{-8} m thick. Low angle x-ray determinations have shown that the structure does not consist of regular layers with constant spacings; a structure which would accommodate the factors which determine stability would be difficult to envision. Further, since the phase is an isotropic liquid, a regularly layered structure is excluded.

One possible structure would consist of irregular forms among which spheres and cylinders with alternatively hydrophobic and hydrophilic outer surfaces are formed according to Figure 8. This suggestion has an immediate attraction; its features are similar to critical phenomena [*cf.*, two-dimensional Ising's model (*40*)]. The resemblance between micellar associations and the fast fluctuation aggregates before phase separations

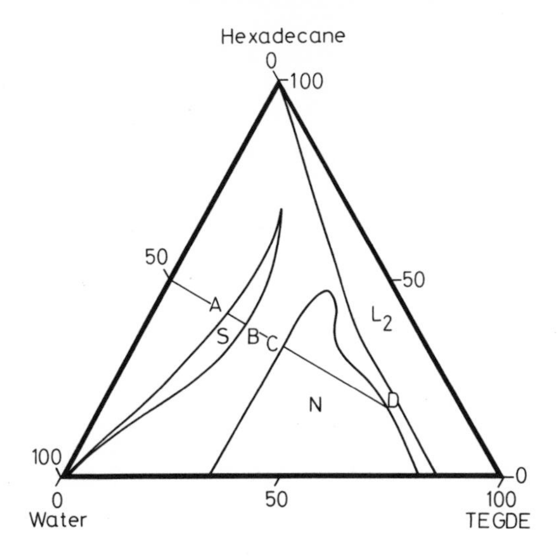

Figure 7. Close to the PIT value two phases with a lamellar structure exist. One of these, the surfactant phase (S), is an isotropic liquid, the other one, (N), is an optically anisotropic liquid crystal with a lamellar structure.

has been noted (*41*). The energy requirements of such a structure would be informative as to its existence.

Since the average interlayer distances are large and since the results of low angle x-ray diffraction exclude a regular layer structure, it appears that the potential energy minimum *vs.* interlayer distance is shallow enough not to influence the structure significantly. Thus, the work stored as potential energy when the layers are bent appears to be an important

Figure 8. The possibility of a non-regular structure of the surfactant phase containing both oil and water dispersed and continuous is attractive but does not appear probable

energy factor. The magnitude of this energy may be estimated by using the compressibility of monomolecular layers assuming the two interfacial tensions ($\gamma_{W/P}$ and $\gamma_{O/N}$) to be equal at a plane interface. The energy demand is calculated from the area difference at bending

$$\Delta E = \int_{A_1}^{A_2} 1/c \cdot dA$$

A_1 and A_2 are areas before and after change, and c is the compressibility

$$c = - 1/\sigma \cdot \partial\sigma/\alpha\Pi$$

in which σ denotes area, and Π denotes interfacial pressure. The quantity $1/c$ has been estimated (*35*), calculated from solubilization data (*36*), or it may be obtained directly from interfacial tension surface area determinations. Reasonable values for $1/c$ are in the range $(0.5–1) \times 10^{-2}$ J/m^2. Assuming a conservatively low value of 10^{-3} J/m^2 the possibility of closed figures according to Figure 8 can be estimated by calculating the energy necessary to form a half-cylinder bridge between two layers. The distance between layers is assumed to be 1.2×10^{-8} m, the length of the half-cylinder joint to be equal to l, and the thickness of the surfactant layer, Δ, to be 2×10^{-9} m. The energy can be estimated as:

$$A_{\text{half-cyl}} = \Pi r l$$

$$dA = \Pi l \cdot dr$$

$$\Delta E = \Pi l/c \cdot \int_{r_1}^{r_1+\Delta} dr$$

$$\Delta E = \Pi l \Delta/c$$

Setting ΔE equal to the thermal energy $kT = 4.14 \times 10^{-21}$ J, a value $l = 6.6 \times 10^{-10}$ m is obtained. This value corresponds to a length of only about two molecules and justifies the comparison with the value kT. The result indicates that closed figures such as those in Figure 8 are less likely.

These results expose two different lamellar structures—S and N in Figure 7—whose only difference is in the amount of emulsifier. The packing conditions of the liquid crystal imply a high degree of order of the CH_2 groups from the one adjacent to the polar group to about two or three groups from the end group (*42, 43, 44*). There is no reason to assume a similar close packing of the surfactant molecules in the surfactant phase, and it seems obvious that the conformation of emulsifier molecules changes when the phase transition goes from the surfactant phase (S) to the lamellar liquid crystalline phase (N). The conformation of the molecules in the surfactant phase has not yet been determined.

Temperature Dependence of the Liquid Crystalline Phase

The presence of a liquid crystalline phase at high surfactant concentrations has been shown by Shinoda (*31*), but the method of presentation renders the evaluation of the temperature dependence of necessary emulsifier concentrations to obtain the liquid crystalline phase difficult. Although several phase diagrams of the system (water, emulsifier, and nonionic surfactant) have been published (*4, 45, 46, 47, 48*), no results have been given on the relation between the surfactant phase and the lamellar liquid crystalline phase in these systems.

The variation of the phase regions with temperature is illustrated in Figure 9 (*38*). At the PIT value (A) the surfactant phase forms an

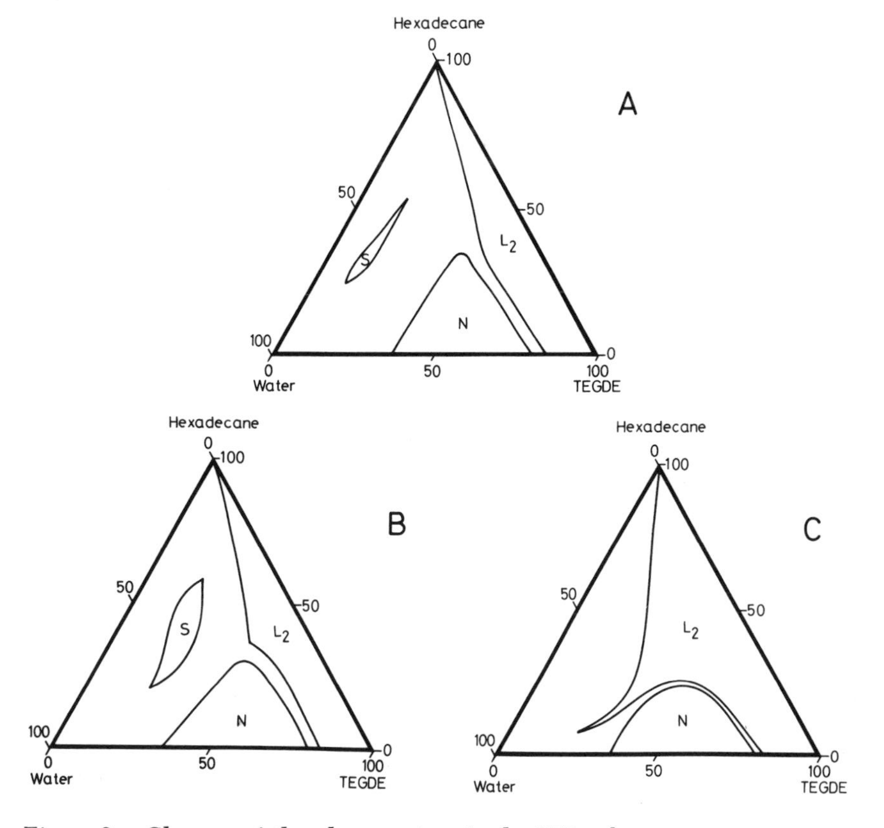

Figure 9. Changes of the phase regions in the PIT value range are more pronounced for the micellar solutions than for the lamellar liquid crystalline phase formed by water and emulsifier

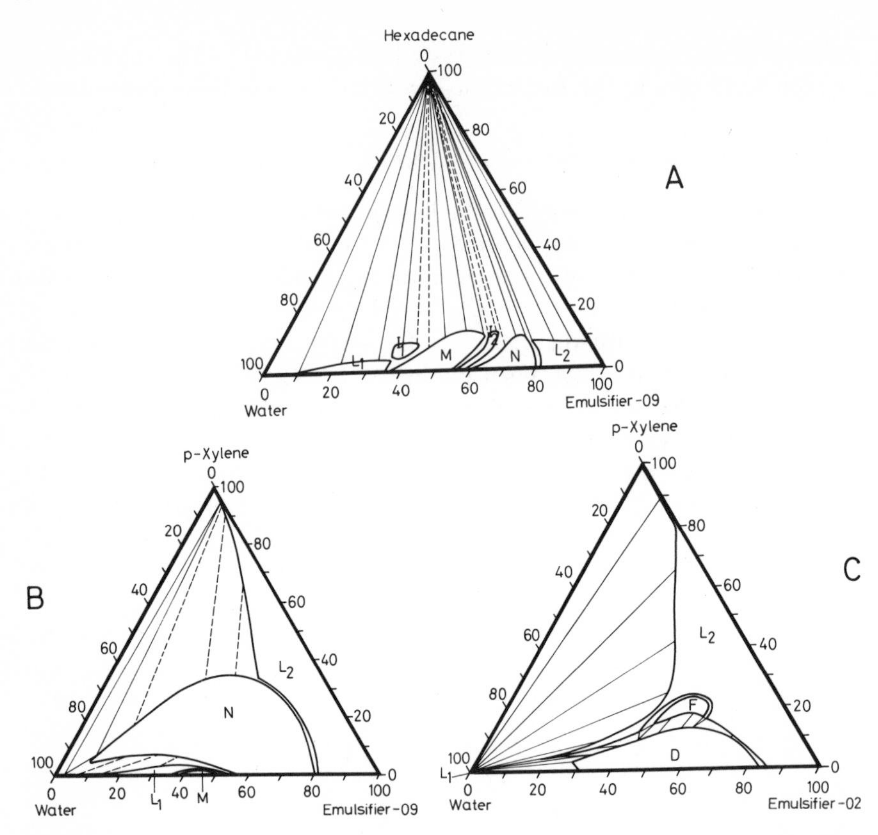

Figure 10. Even complex phase equilibria follow the general trends observed for simpler systems in the PIT range (cf. Figure 9). Emulsifiers are polyoxyethylene nonyl phenol ethers.

isolated region with fairly constant emulsifier concentration. Higher concentrations induce the liquid crystalline phase. Increased temperature will transfer the surfactant phase region to a composition containing higher amounts of emulsifier. At higher temperatures the surfactant phase coalesces with the cosolvency region L_2, forming a narrow solubility region toward the water corner. These drastic changes of the surfactant phase region contrast sharply with the conditions at low hydrocarbon content. The maximum solubility of water in the liquid emulsifier and the emulsifier range for obtaining the liquid crystalline phase are only slightly altered in the temperature range in Figure 9.

According to the present results the marked changes of isotropic liquid regions at the PIT described by Shinoda and co-workers (*29, 30, 31, 32, 33*) appear limited to micellar associations and solubilization. The cosolvency of water and hydrocarbon in the emulsifier and the liquid crystalline phase region seem to undergo only small changes.

Phase Regions at Constant Temperature

Figure 1 shows how a change of hydrocarbon can affect the phase regions. Figure 10 illustrates the influence of emulsifier properties. A reduction of the mean polyoxyethylene chain length from 8.6 to 4 drastically changed the phase regions.

Both results agree with the general trend of the variations expected when the PIT is considered. At first (Figure 10A, B) a change in hydrocarbon from aliphatic to aromatic reduces the PIT by about 60°C (*30*). Figure 10A represents phase regions far below the PIT; here the solubility is limited to the aqueous phase; Figure 10B is observed at a temperature near the PIT (*32*). The solubility in both liquids and the equal importance of these in governing the phase regions are noticeable. A reduction of the polyoxyethylene chain from an average of 8.6 to about 4 should further reduce the PIT by about 20°C; Figure 10C illustrates phase conditions at temperatures well above PIT; the narrow solubility region of the L_2 area towards the water corner is a conspicuous feature (*cf.* Figure 9).

The behavior of a series of polyoxyethylene alkyl ether nonionic surfactants is also illustrative. According to Figure 11 the dioxyethylene (A) compound does not form liquid crystals when combined with water. Its solutions with decane dissolve water only in proportion to the amount of emulsifier. The tetraoxyethylene dodecyl ether (B) forms a lamellar liquid crystalline phase and is not soluble in water but is completely miscible with the hydrocarbon. The octaoxyethylene compound (C) is soluble in both water and in hydrocarbon and gives rise to three different liquid crystals: a middle phase, an isotropic liquid crystal, and a lamellar phase containing less water. If the hydrocarbon *p*-xylene is replaced by hexadecane (D), a surfactant phase (L) and a lamellar phase containing higher amounts of hydrocarbon are formed in combination with the tetraoxyethylene compound (B-D).

The results when the hydrocarbon is varied in combination with water and tetraoxyethylene dodecyl ether are within the pattern expected from the PIT concept. Combination with the aromatic hydrocarbon should be characterized by a low PIT, and the diagram displays a system above the PIT, while the regions containing hexadecane should possess features of a system below the PIT value. Figure 11B, D confirms this prediction—*i.e.*, 11B is the type of diagram expected at temperatures well above the PIT; on the other hand, Figure 11D is distinctly below the PIT but not far from it.

Increasing polyoxyethylene chain length (Figure 11 A-C) increases the PIT, similar to an increase in surfactant cloud point. The general trend expected (such as increasing water solubility) is also observed in

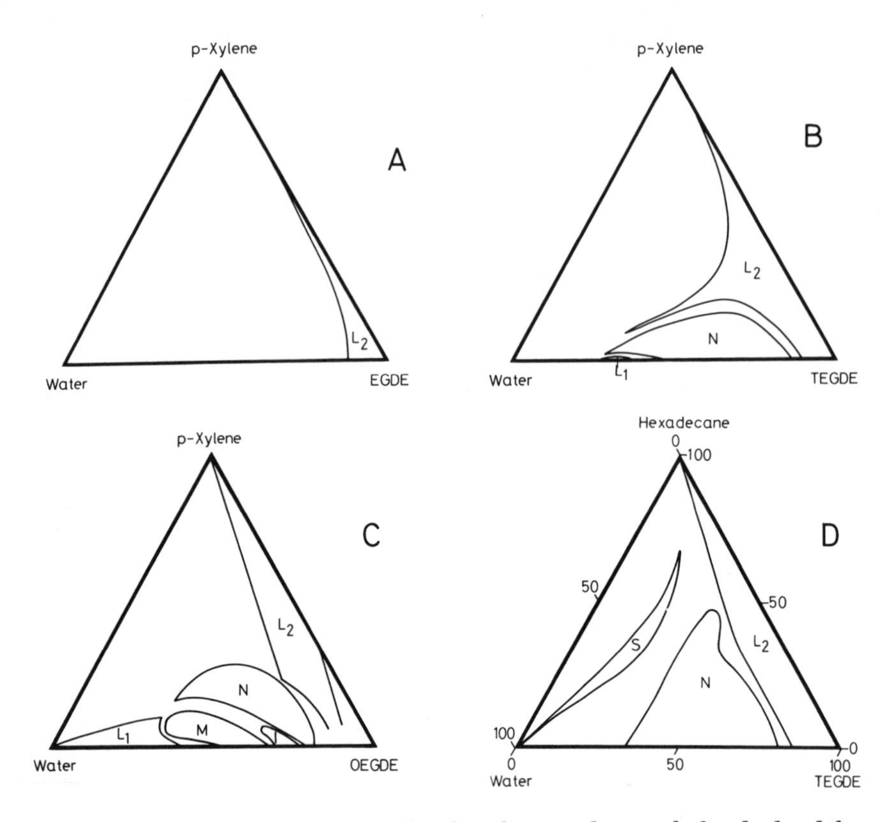

Figure 11. Increasing the PIT value by altering the emulsifier hydrophilic group (A-C) or nature of the hydrocarbon (B-D) give changes expected from the PIT concept (cf. *Figure 9*)

OEGDE = octaoxyethylene dodecyl ether
EGDE = oxyethylene dodecyl ether
TEGDE = *tetraoxyethylene dodecyl ether*

the diagrams. In addition the increased volume of the hydrophilic part of the emulsifier gives rise to the formation of a middle phase and one isotropic liquid crystal in addition to the lamellar structure.

The temperature dependent alterations of the equilibria between different phases have not yet been studied. Concerning the results in Figure 9, the changes along the water–emulsifier axis should be less pronounced than shifts at compositions with higher amounts of hydrocarbon.

Conclusions

Regions containing a liquid crystalline phase in systems comprised of water, a hydrocarbon, and a nonionic emulsifier (polyoxyethylene alkyl ether) are affected by the temperature-induced changes characteristic of the variations brought about in the isotropic liquid phases. These are most distinctive in the phase inversion temperature (PIT) range.

The liquid crystalline regions obey the general rules for the liquid phases, but only where the hydrocarbon content is high. Along the water–emulsifier axis the changes with temperature are small in the PIT range; this indicates that the structure of the liquid crystalline phase depends mainly on short range emulsifier–water interactions, which limits the solubility of water into the emulsifier.

The preliminary results reported here indicate that the general changes induced in the PIT range may help to systematize the complex phase behavior of nonionic surfactants when they are combined with water and hydrocarbons.

Literature Cited

1. Gray, G. W., Winsor, P. A., ADVAN. CHEM. SER. (1976) **152,** 1.
2. Danielsson, I., ADVAN. CHEM. SER. (1976) **152,** 13.
3. Shinoda, K., Nakagawa, T., Tamamushi, B. J., Isemura, T., "Colloidal Surfactants," K. Shinoda, Ed., Academic Press, New York, 1963.
4. Friberg, S., Mandell, L., Fontell, K., *Acta Chem. Scand.* (1969) **23,** 1055.
5. Schick, M., *J. Phys. Chem.* (1963) **67,** 1796.
6. Ray, A., Némethy, G., *J. Phys. Chem.* (1971) **75,** 809.
7. Bailey Jr., F. E., Collard, R. V., *J. Appl. Polymer Sci.* (1959) **1,** 56.
8. Davies, J. T., *Proc. Int. Congr. Surf. Activity, 2nd* (1957) **1,** 426.
9. Bailey Jr., F. E., Koleske, J. V., "Nonionic Surfactants," M. Schick, Ed., Chap. 23, M. Dekker, New York, 1967.
10. Jones, E. R., Berry, C. R., *Phil. Mag.* (1927) **4,** 841.
11. Stainsby, G., Alexander, A. E., *Trans. Faraday Soc.* (1950) **46,** 587.
12. Aranov, R. H., *J. Phys. Chem.* (1963) **67,** 556.
13. Vold, M. J., *J. Colloid Sci.* (1950) **5,** 506.
14. Hill, T. L., "Thermodynamics of Small Systems," Vols. I and II, W. J. Benjamin, New York, 1963, 1964.
15. Corkill, J. M., Goodman, J. F., Walker, T., *Trans. Faraday Soc.* (1967) **63,** 759.
16. Hall, D. G., Pethica, B., "Nonionic Surfactants," M. Schick, Ed., Chap. 16, M. Dekker, New York, 1967.
17. Corkill, J. M., Goodman, J. F., Tate, J. R., "Hydrogen-Bonded Solvent Systems," A. K. Conington and P. Jones, Eds., Taylor & Francis, London, 1968.
18. Becher, P., "Nonionic Surfactants," M. Schick, Ed., Chap. 15, M. Dekker, New York, 1967.
19. Corkill, J. M., Walker, T., *J. Colloid Interface Sci.* (1972) **39,** 621.
20. Clarke, D. E., Hall, D. G., *Kolloid-Z. Z. Polym.* (1972) **250,** 961.
21. Schick, M. J., Gilbert, A. H., *J. Colloid Sci.* (1965) **20,** 464.
22. Becher, P., *J. Colloid Sci.* (1965) **20,** 728.
23. Gratzer, W. B., Beaven, G. H., *J. Phys. Chem.* (1969) **73,** 2270.
24. Shinoda, K., Nakagawa, T., Tamamushi, B. J., Isemura, T., "Colloid Surfactants," K. Shinoda, Ed., Chap. 1, Academic Press, New York, 1963.
25. Shinoda, K., Ogawa, T., *J. Colloid Interface Sci.* (1967) **24,** 56.
26. Kitahara, A., Ishikowa, T., Tanimori, S., *J. Colloid Interface Sci.* (1967) **23,** 243.

27. Kon-no, K., Kitahara, A., *J. Colloid Interface Sci.* (1971) **35**, 636.
28. *Ibid.*, (1971) **37**, 469.
29. Saito, H., Shinoda, K., *J. Colloid Interface Sci.* (1967) **24**, 10.
30. Shinoda, K., Arai, H., *J. Phys. Chem.* (1964) **68**, 3485.
31. Shinoda, K., Saito, H., *J. Colloid Interface Sci.* (1968) **26**, 70.
32. Saito, H., Shinoda, K., *J. Colloid Interface Sci.* (1970) **32**, 647.
33. Shinoda, K., Kunieda, H., *J. Colloid Interface Sci.* (1973) **42**, 381.
34. Prince, L. M., *J. Colloid Interface Sci.* (1965) **23**, 165.
35. *Ibid.*, (1968) **29**, 216.
36. Robbins, M., "The Theory of Microemulsions," *AIChE Meetg., Tulsa, Okla.*, March 1974.
37. Friberg, S., Lapczynska, I., *Colloids Polymers*, in press.
38. Doi, T., unpublished data.
39. Lange, H., "Nonionic Surfactants," M. Schick, Ed., Chap. 14, M. Dekker, New York, 1967.
40. Ogita, N. *et al.*, *J. Phys. Soc. Japan* (1969) **26S** ,145.
41. Shinoda, K., Friberg, S., *Adv. Surface Colloid Sci.*, in press.
42. Charvolin, J., Manneville, P., Deloche, B., *Chem. Phys. Lett.* (1973) **23**, 345.
43. de Gennes, P. G., *Phys. Lett.* (1974) **47A**, 123.
44. Marćelja, S., *Nature* (1973) **241**, 451.
45. Salisbury, R., Leuallen, E. E., Chavkin, L. T., *J. Amer. Pharm. Assoc., Sci. Ed.* (1954) **43**, 117.
46. Burt, B. W., *J. Soc. Cosmetic Chem.* (1965) **16**, 465.
47. Lachampt, F., Vila, R. M., *Amer. Perf. Cosm.* (1967) **82**, 29.
48. Groves, M. J., Mustafa, R. M. A., Carles, J. E., *J. Pharm. Pharmac.* (1974) **26**, 616.

RECEIVED November 19, 1974.

4

Liquid Crystalline Phases in Biological Model Systems

K. LARSSON and I. LUNDSTRÖM[1]

University of Göteborg and Chalmer's University of Technology, Fack, S-402 20 Göteborg 5, Sweden

*The present knowledge about molecular organization in lyotropic liquid crystalline phases is summarized. Particular attention is given to biologicaly relevant structures in lipid–water systems and to lipid–protein interactions. New findings are presented on stable phases (gel type) that have ordered lipid layers and high water content. Furthermore, electrical properties of various lipid structures are discussed. A simple model of 1/*f *noise in nerve membranes is presented as an example of interaction between structural and electrical properties of lipids and lipid–protein complexes.*

The significance of liquid crystals in biological systems is obvious from the fact that most life processes require molecular disorder and mobility with maintenance of orientation of the functional groups involved; this is just the molecular organization that is characteristic of the liquid crystalline state. In a few cases, for example in muscles and in the nerve myelin sheath, there are liquid crystalline regions with three-dimensional extension. More often, however, the structure is two-dimensional, consisting of a unit layer of a lamellar liquid crystalline phase. The membrane that covers all cells and cell organelles has a molecular arrangement of this type. Since there are very few physical methods for investigating two-dimensional structures in an aqueous environment, the use of related liquid crystalline phases as models provides an important basis for our understanding of membrane structure.

Lyotropic liquid crystalline phases had been utilized technically for a long time before their structures were known. A milestone in the elucidation of their structure was the introduction of the liquid chain concept. In 1958, on the basis of evidence from IR spectroscopy, Chapman pro-

[1] Present address: Chemical Center, Box 740, S-22007 Lund, Sweden

posed that a high temperature phase transition in anhydrous soaps was caused by melting of the chains (*1*). A few years later, the complete structure of the most common liquid crystalline phases in soap–water systems was revealed by the pioneering work of Luzzati and co-workers (*2*).

A characteristic feature of molecules that form lyotropic liquid crystals is their surface activity. Because of the amphiphilic nature of the molecules, they orient upon contact with solvent molecules, giving rise to polar and nonpolar regions that are separated by the polar end groups. All structures known fit one of those made possible by the various curvatures of the interface between two liquid regions, with molecular size taken into consideration. It is therefore not surprising that the earlier treatment of the structure of lyotropic liquid crysals was unsuccessful since the molecules were regarded as stiff rods.

The lyotropic liquid crystalline phases relevant to biological systems consist of water and lipids and usually proteins also. The lipids listed in Table I occur in cell membranes; all form liquid crystalline phases with water.

Different biological tissues are complex mixtures of various lipid types, and each one is usually represented by numerous homologs with

Table I. Lipids in Cell Membranes

Lipid	*Typical Example*	*Formula*
Phospholipids	dipalmitoyllecithin	$CH_2—OCO(CH_2)_{14}CH_3$ | $CH_2—OCO—(CH_2)_{14}CH_3$ | $CH_2—OPO_2^-O(CH_2)_2N^+(CH_3)_3$
Sphingolipids	cerebroside	$CH_3(CH_2)_{12}—C(H)=C(H)—C(H)(OH)—C(H)(NH—C(=O)—R)—CH_2—O—CH—$ $—CH—HCOH—HOCH—HOCH—HC—CH_2OH$ (ring closed through —O— between first CH and HC)
Glycerides	1-monopalmitin	CH_2OH | $CHOH$ | $CH_2OCO(CH_2)_{14}CH_3$

different fatty acid composition. A remarkable property of such mixtures is that they usually behave as a single component with respect to the phase rule. As is discussed below, lipid phase transitions are of functional importance in membranes, and the possibility of varying the fatty acid pattern provides a way to vary the phase transition temperature as well as the transition range.

Structures in Lipid–Water Systems

The three fundamental lyotropic liquid crystal structures are depicted in Figure 1. The lamellar structure with bimolecular lipid layers separated by water layers (Figure 1, center) is a relevant model for many biological interfaces. Despite the disorder in the polar region and in the hydrocarbon chain layers, which spectroscopy reveals are close to the liquid states, there is a perfect repetition in the direction perpendicular to the layers. Because of this one-dimensional periodicity, the thicknesses of the lipid and water layers and the cross-section area per lipid molecule can be derived directly from x-ray diffraction data.

The hexagonal arrangement of cylinders formed by lipid molecules in a continuous water medium (Figure 1, right) is observed in all systems with liquid crystals, provided that there is a micellar state at high water content. Only a few lipids which have a large polar head group relative to the hydrocarbon chain portion of the molecule show this phase, *e.g.* lysolecithin (obtained from lecithin when one chain is removed) and psychosine (related to cerebroside in the same way). The x-ray diffraction pattern of this phase can be indexed as a two-dimensional hexagonal lattice, and one may apply simple geometry in order to determine the radius of the lipid cylinders, the distance between adjacent cylinders, and the area per lipid molecule at the water interface.

The inversed hexagonal structure, with water cylinders arranged in a matrix formed by the disordered hydrocarbon chains (Figure 1, left), is a common structure in aqueous systems of lipids of biological origin. There is usually no problem in determining the true alternative between the two hexagonal structures from the x-ray data, and the molecular dimensions can then be calculated. The occurrence of this structure in complex lipids results from the molecular shape; two hydrocarbon chains are usually

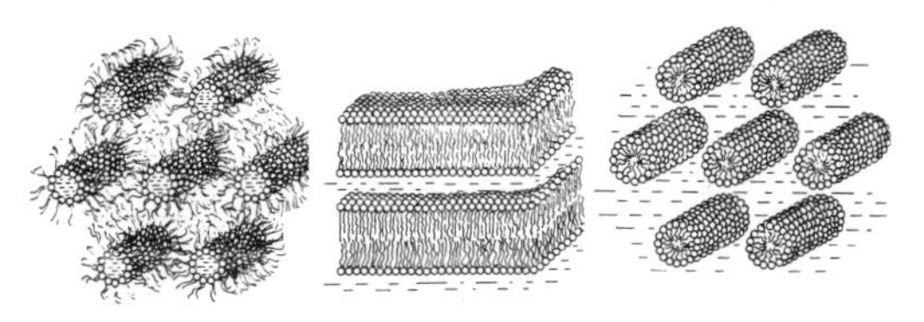

Figure 1. Schematic of the three most common phases that occur in aqueous lipid systems

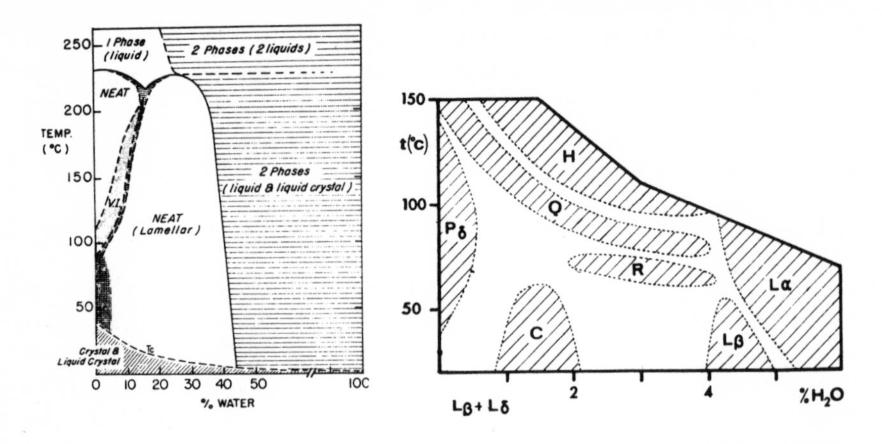

Journal of Lipid Research and Nature

Figure 2. Egg lecithin. Left: phase diagram (3), and right: details of the region with low water content (4)

linked to the polar end group. If the interface has a curvature so that the chains diverge, they will have more space than in the lamellar structure. This structure often forms from the lamellar one upon heating. It is known from spectroscopic studies that chain mobility increases toward the methyl end. This hexagonal structure is also favorable for such thermal movement, and the transition from the lamellar phase when thermal movement increases is therefore not surprising.

One type of lipid that is dominant in biological interfaces is lecithin, and lecithin–water systems have therefore been examined extensively by different physical techniques. Small's binary system (3) for egg lecithin–water is presented in Figure 2. The lamellar phase is formed over a large composition range, and, at very low water content, the phase behavior is quite complex. Their structures as proposed by Luzzati and co-workers (4) are either lamellar with different hydrocarbon chain packings or based on rods; both types are discussed below.

The swelling of lecithin (and of all other neutral lipids studied) proceeds until a water layer thickness of about 20 A is reached. Beyond this limit, the region with higher water content is often discribed as consisting of two phases, water and the lamellar phase. This is not correct, however, since concentric structures are formed in this whole range. The spherical so-called liposomes, with lipid layers alternating with water layers, are not restricted only to very dilute systems, but, together with the cylindrical arrangement of concentric layers, they represent a particular structural state above the limit of water swelling. Our knowledge of the equilibrium state of these dispersions and variations in shape and size of the colloidal particles is very limited even though liposomes are used frequently as membrane models. It is possible to break these particles mechanically, for example by ultrasound, and a large proportion of particles consisting of a single bilayer is then obtained. These so-called vesicles can be separated, and they have obvious advantages over liposomes as membrane models.

The phase behavior of a synthetic lecithin, dipalmitoyllecithin, as analyzed by Chapman and co-workers (*5*), is diagrammed in Figure 3. The main features are the same as in the phase diagram of egg lecithin: a mixture of numerous homologs. As a consequence of the variation in fatty acid chain length, the chain melting point is lowered which means that the critical temperature for formation of liquid crystalline phases is reduced. This temperature is about 42°C for dipalmitoyllecithin, and, if the lamellar liquid crystal is cooled below this temperature, a so-called gel phase is formed. The hydrocarbon chains in the lipid bilayers of this phase are extended, and they can be regarded as crystalline. The gel phase and the transitions between ordered and disordered chains are considered separately.

Cubic phases have also been observed in lipid–water systems. Such a phase is not a true liquid crystal since it exhibits three-dimensional periodicity, but all physical properties are closely related to those of the lamellar and hexagonal phases. For the structure derived for the cubic phase of strontium myristate (*6*), *see* Figure 4. The lipid molecules are joined so that the polar groups form rods, and these rods are joined into two separate three-dimensional networks. The lipid region as well as the polar region in this structure form continuous media. There is also evidence for a cubic structure with closed water aggregates. It is formed in monoglycerides of medium chain length when the lamellar phase is heated

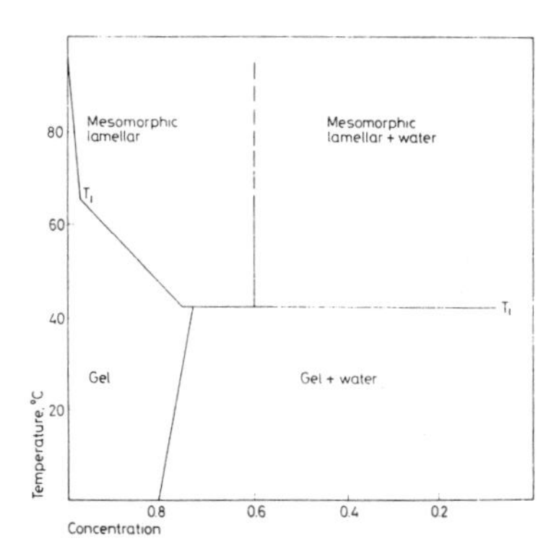

Chemistry and Physics of Lipids

Figure 3. Phase diagram of the dipalmitoyllecithin–water system (5)

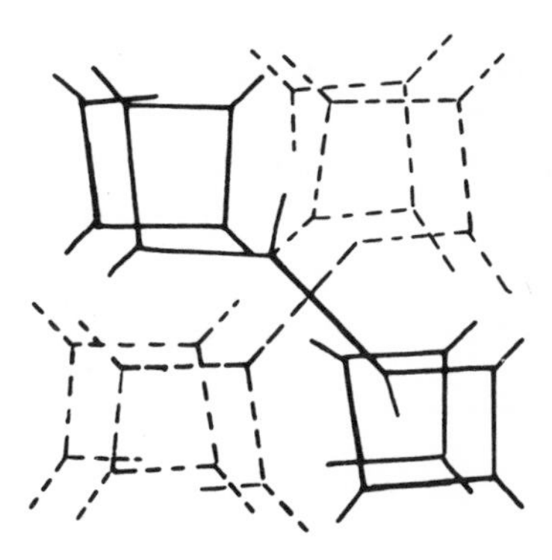

Nature

Figure 4. The rod systems formed by the polar groups in the cubic phase that forms in strontium myristate (6)

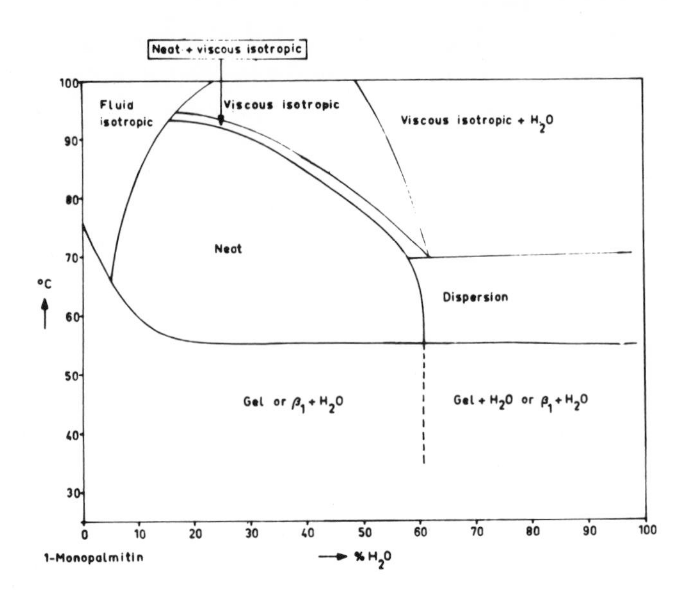

Figure 5. The binary system of monopalmitin–water

(Figure 5), and, with longer chains ($> C_{20}$), it transforms into the hexagonal structure with closed water cylinders upon further heating. Structures based on space-filling polyhedra were analyzed (7), and the x-ray data are in good agreement with a structure consisting of a body-centered arrangement of polyhedra with faces formed by six squares and eight hexagons (*see* Figure 6). This structure was later confirmed by freeze-etching electron microscopy (8).

Aqueous systems of molecules as different as lecithins and monoglycerides have very similar phase diagrams (*cf.* Figures 3 and 5), which illustrates that lipids with similar size relations between hydrophobic and hydrophilic regions (expressed for example by the HLB value) give the same type of water interaction. If ionic groups are present, the lamellar

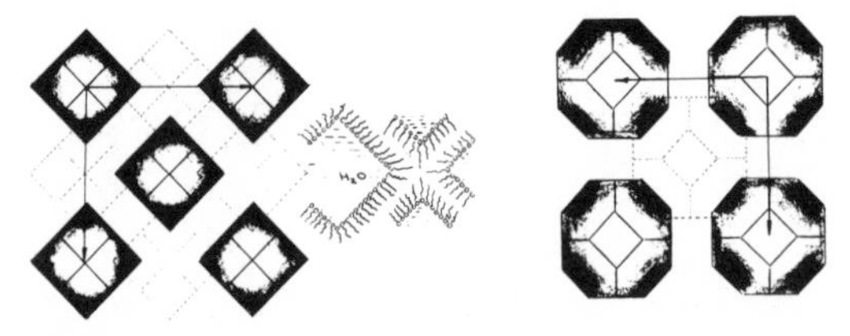

Figure 6. Cubic structures in lipid–water systems based on space-filling polyhedra. The data from the monoglyceride–water cubic phases fit with the body-centered structure to the right.

phase swells to a higher maximum water content, but, besides this, there are no major differences in the behavior of ionic lipids.

Multicomponent systems, which contain different types of lipids and water, have the same phases as were described above. The phase diagrams are of course complicated by the coexistence of many phases in equilibrium, but, apart from that, the same relations between structure and the size–shape relations of the hydrophobic and hydrophilic regions of the lipid mixtures as in the binary systems seem to be valid. Model systems related to different diseases in which liquid crystalline phases are involved were studied, particularly by Small. As an example, a ternary system in which the most important components of atheriosclerotic lesions are involved (*10*) is illustrated in Figure 7. The addition of phospholipids and water to the phase diagrams gives a physical basis for understanding the formation of these pathological lipid deposits (*11*). The formation of gall stones by cholesterol precipitation in the bile was examined in a similar way (*12*).

Lipid–Protein Interaction

Although the association between lipids and proteins is fundamental in understanding the physiological functions of membranes, information on such structures is very limited. Studies of a few systems of lipids and globular proteins indicate that the proteins tend to remain in their native form. The structures can be separated into two somewhat simplified types. Usually the lipid structure seems to dominate, and the protein molecules are incorporated into liquid crystalline structures of lipids. In other cases, the lipid molecules are distributed within the protein units,

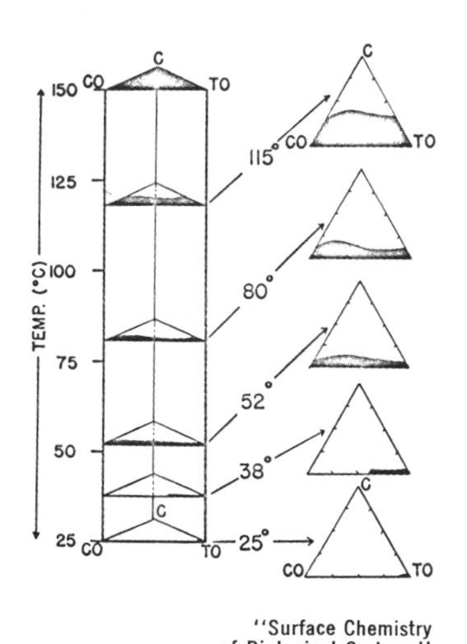

Figure 7. The ternary system cholesteryl oleate–cholesterol–triolein (CO–C–TO) at different temperatures (10). The darkened region corresponds to one isotropic phase whereas the remainder consists of two or three phases.

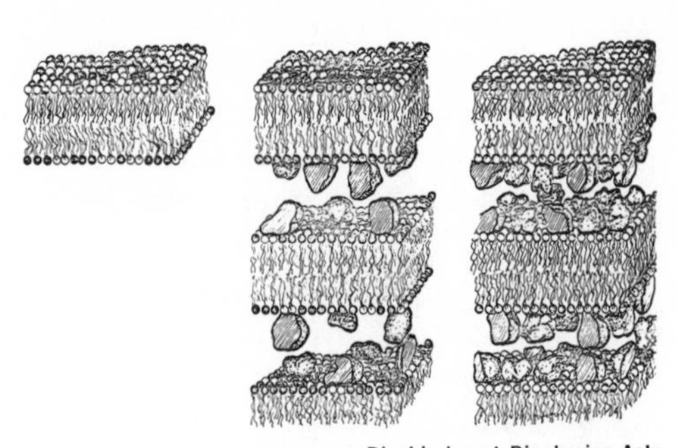

Biochimica et Biophysica Acta

Figure 8. Molecular arrangements in aqueous precipitates of insulin and lecithin–cardiolipin (13). Left: the lipid bilayer structure used in the preparations; center: the precipitate formed by a 9/1 molar ratio of lecithin–cardiolipin; and right: the precipitate formed by a 6/4 molar ratio.

and it then might be possible to crystallize the complex and to obtain a detailed structure determination.

An experimental complication is the difficulty in effecting molecular interaction between the components. The usual technique for preparing lipid–protein phases in an aqueous environment is to use components of opposite charge. This in turn means that the lipid should be added to the protein in order to obtain a homogeneous complex since a complex separates when a certain critical hydrophobicity is reached. If the precipitate is prepared in the opposite way, the composition of the complex can vary since initially the protein molecule can take up as many lipid molecules as its net charge, and this number can decrease successively with reduction in available lipid molecules. It is thus not possible to prepare lipid–protein–water mixtures, as in the case of other ternary systems, and to wait for equilibrium. Systems were prepared that consisted of lecithin–cardiolipin (L/CL) mixtures with (a) a hydrophobic protein, insulin, and with (b) a protein with high water solubility, bovine serum albumin (BSA).

In the insulin–L/CL complexes, it is evident from the dimensions of the lamellar liquid crystalline phase that insulin is simply associated electrostatically with the L/CL bilayer and that it replaces water (*13*). The amounts depend on the number of charges (*see* Figure 8). The limit of protein association is reached when no more surface is available at the bilayer–water interface.

BSA–L/CL complexes are also lamellar liquid crystals. There are two alternative models (Figure 9) which can explain the observed lattice

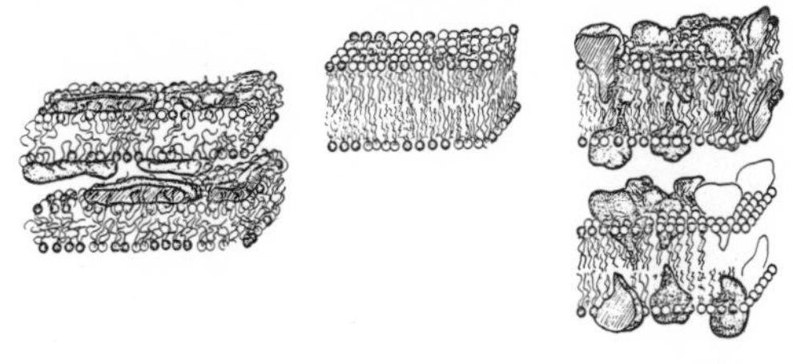

Biochemistry

Figure 9. Molecular arrangements in aqueous precipitates of bovine serum albumin and lecithin–cardiolipin (14). Center: lipid bilayers; and left and right: two alternative structures of the precipitates based on the x-ray diffraction spacings.

dimensions (*14*). In the most probable alternative, parts of the protein molecules penetrate the lipid layer; consequently, there is a hydrophobic lipid–protein interaction. The other alternative is that the thickness of the lipid bilayer is so reduced that there are direct contacts between water and hydrocarbon chains. All available data on lamellar lipid–water liquid crystals reveal similar values for the cross-section area per hydrocarbon chain at the bilayer–water interface that are quite different from those for this second structural alternative. The probable mechanism for formation of the complex is therefore an initial electrostatic effect that brings the protein and lipid molecules together; subsequently, a structure with optimum short-range interaction between the components is adopted.

The association of lipids with proteins in dilute aqueous solutions was studied by Tanford (*15*). He identified different types of interaction that depend on the number of associated lipid molecules. He also analyzed the relation between lipid association into micelles and the competing binding of lipids to proteins.

Observations were made of lipid–protein phases in which the structure is determined mainly by the protein. Raman spectroscopy is a useful method for structure analysis of such phases. The structures described above were analyzed successfully by an x-ray diffraction technique. Lipid–protein complexes, however, are often amorphous, and alternative methods to study their structures are therefore needed. It was demonstrated that Raman spectroscopy can be used to obtain structural information about lipid–protein interaction (*16, 17*). It is thus possible to determine the conformation as well as the type of environment of the lipid molecules. With the protein, interpretation is more complicated. It is usually possible to determine whether the complex has the same protein conformation as the component used in the preparation, or, if a change occurs, it may be possible to correlate it with denaturation of the pure protein. For complexes formed by long-chain alkyl phosphates and insu-

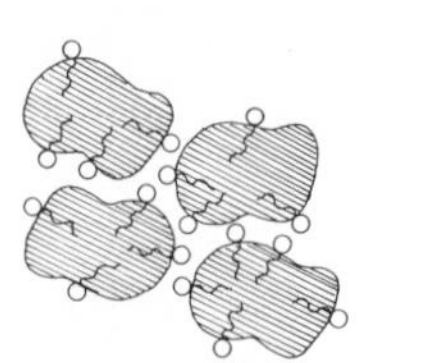
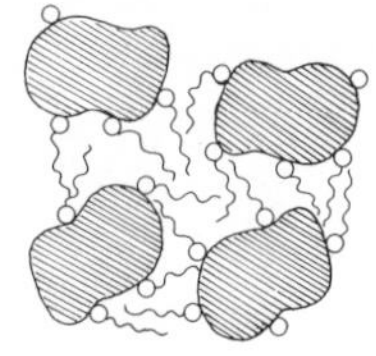

Figure 10. Models of complexes between long-chain phosphate esters (sodium salts of the monoesters) and insulin based on Raman spectroscopy. Shorter chains (e.g. C_{10}) have a protein environment whereas lipid regions with disordered chains are formed when chains are longer ($>C_{14}$).

lin, the structural features in Figure 10 were derived from the Raman spectra. Insulin remains in its native form in complexes formed by phosphates. In the members with long chains ($> C_{14}$), the hydrocarbon chains are surrounded by other chains which indicates that they form the usual type of hydrocarbon chain regions. When chain lengths are short, ($< C_{10}$), however, the phosphate ester molecules have a different and more polar environment, and they are therefore likely to be distributed in hydrophobic pockets of the insulin molecules. The hydrocarbon chains in both structure types are disordered and are mainly of gauche conformation. On drying, the components separate and the chains crystallize. An important role of water in lipid–protein complexes is probably to function as a space-filling polar medium.

The possibility of obtaining information about lipid–protein interaction makes Raman spectroscopy a useful technique for structural studies of membranes. As an illustration of spectra recorded from biological samples, *see* the Raman spectrum of a frog sciatic nerve in Figure 11. The C–H stretching vibration region is characteristic of lipid bilayers in a

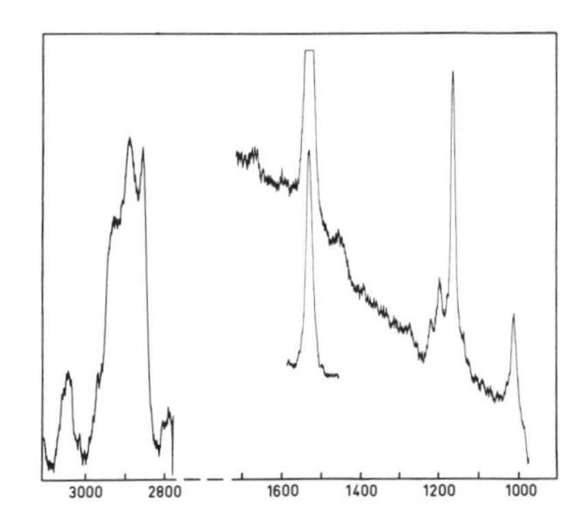

Figure 11. Raman spectrum of a frog sciatic nerve in Ringer solution

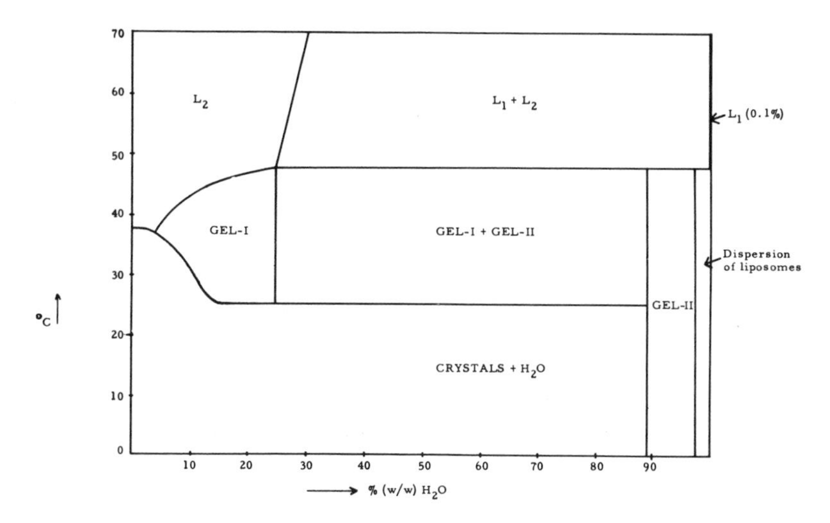

Figure 12. The binary system tetradecylamine–water

mainly disordered state, which agrees with findings from structural studies of this particular multilamellar structure by numerous other techniques.

The Gel State in Lipid–Water Systems

Lipid–water gel phases were previously regarded as metastable structures that are formed before separation of water and lipid crystals when the corresponding lamellar liquid crystal is cooled. New information on gel phases (*see* below) reveals that they can form thermodynamically stable phases with very special structural properties. This characteristic makes them as interesting as the lamellar liquid crystals from a biological point of view.

The binary system of tetradecylamine–water is diagrammed in Figure 12. There are two gel phases with different water layer thicknesses but with the same bilayer structure (*18*). The molecules are extended and vertical in the bilayer in both forms, and the existence of two forms can be explained as follows. The gel-I phase swells to a water layer thickness of about 14 A, which is about the same as that observed in all known lamellar liquid crystals of neutral molecules (some of which can be cooled to give a gel phase). At high water concentrations, there are enough water molecules to ionize a certain proportion of the amine groups, and, at a critical concentration, the electrostatic repulsion of the bilayers will balance the van der Waals attractive forces. When the proton concentration of the water present was changed, this critical concentration shifted as expected from the changed charge density.

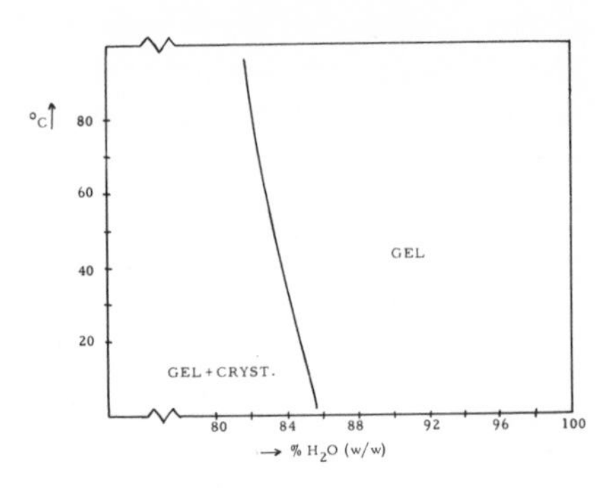

Figure 13. The binary system cholesterol dihydrogenphosphate–water

Cholesterol sulfate and cholesterol monophosphate both form aqueous phases of the liquid crystal and gel type. This is remarkable since it is believed that the occurrence of lyotropic liquid crystals requires a hydrocarbon region formed by flexible chains in a liquidlike state. The sulfate was so unstable chemically that it was impossible to obtain phase equilibria at high water content, but the general behavior was the same as that of the dihydrogen monophosphate. The phase diagram of the system cholesterol dihydrogenphosphate–water is presented in Figure 13 (*19*). The crystals exist both in anhydrous form and as a hydrate. Aqueous phases with water exist only at very high water contents. Above about 85 wt % water, a stable gel phase is formed; it exists as a very viscous homogeneous phase up to a water content of 99 wt %. A transition is observed when the gel phase is heated; from x-ray wide-angle diffraction studies, this means that there is increased disorder in the lipid bilayers. In analogy with other lipid–water systems, this transition is therefore described as a gel → liquid crystal transition. Also, in this case it is possible to explain the minimum water layer thickness required for formation of the gel phase by a critical electrostatic repulsion that results from ionization of some phosphate groups.

The lamellar spacing of a monoglyceride gel phase as a function of water content is plotted in Figure 14. The gel phase of the neutral monoglyceride has a lipid bilayer thickness of 49.5 A, and it swells to a unit layer thickness of 64 A (*20*). If an ionic amphiphilic substance (*e.g.* a soap) is solubilized in the lipid bilayer, it is possible to obtain a gel phase with high water content. As with the gel phases with infinite swelling that were discussed above, there is, however, a minimum water layer thickness which in this monoglyceride gel is about 40 A.

The existence of a forbidden water layer thickness range, which seems to be a general phenomenon with these gel phases, might be relevant to cell adhesion and equilibrium distances at cell contact. The gel represents one type of lipid bilayer structure that occurs in membranes (*see* below), and, because of the dominance of neutral lipid molecules, the charge density can be similar to that of the gel phases described here. Hypothetical cells with the same surface structure as these gel phases obviously cannot come closer to each other than the limit set by the minimum water layer thickness unless the surface structure is changed. The presence of counterions in the water medium of the gel phases, however, has a drastic effect on these distances, and their effect on cell contact distances is also well recognized. As little as 0.3 wt % sodium chloride in the aqueous medium is enough to prevent swelling of the monoglyceride gel with solubilized ionic amphiphiles above a water layer thickness of

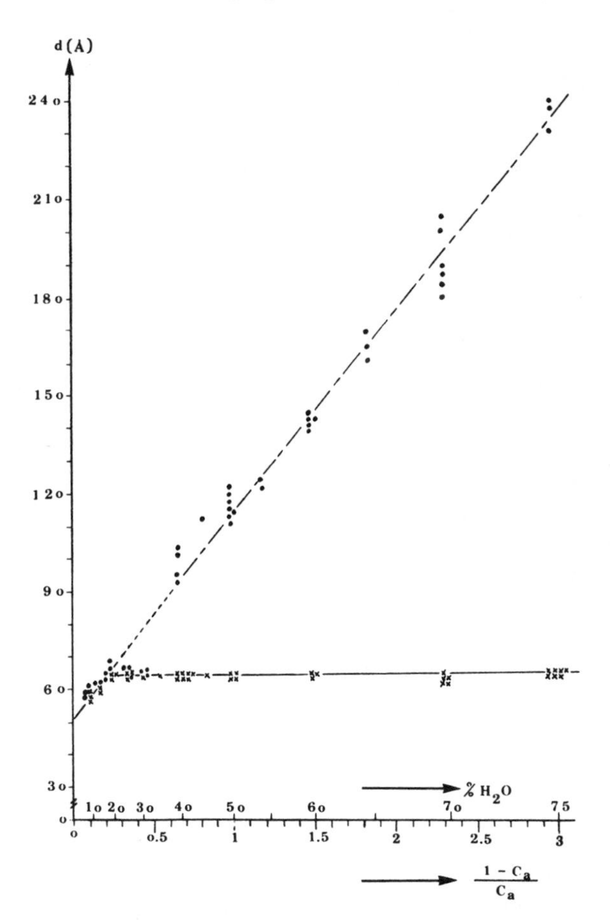

Figure 14. X-ray diffraction data for the gel phase of a monostearin sample at 25°C. ×—× at pH 5–6, the limited swelling of a monoglyceride gel; and •---•: at pH 7, the swelling up to high water content in the presence of charged groups (sodium stearate/monoglyceride molecular ratio, 1/60).

14 A, and, consequently, there is no minimum water layer thickness. If the salt concentration is increased above 2 wt %, it is not possible to get any water at all between the lipid bilayers.

Membranes and the Significance of the Hydrocarbon Chain Structure

Although numerous models for the structure of membranes have been proposed, the structural features which are generally accepted at present are rather similar to the original Danielli–Davson model. There is convincing evidence that the structure is dominated by lipid bilayers. The state of order of the hydrocarbon chains is now being studied extensively by many groups (*see* below). Less is known about the proteins. Besides the proteins that are located on the outside according to the Danielli–Davson model, there are also proteins that are partly buried in the hydrophobic interior of the lipid layer; however, little is known about the lipid–protein interaction.

The transition between crystalline and melted hydrocarbon chains in membranes was studied calorimetrically, and the possibility of varying the fatty acid composition and therefore the phase transition temperature in certain microorganisms has provided valuable information on such transitions. It is known that the thermal transition between the lamellar liquid crystalline phase and the gel phase in aqueous systems has a correspondence in liposomes, vesicles, and even membranes. This transition is described as a transition between liquid and crystalline chains. It should be noted, however, that the chains are usually not truly crystalline in the gel state. The occurrence of a single x-ray short spacing at 4.15 A, which is used for identification, shows that the chains are arranged in a hexagonal structure with rotational or oscillational disorder of the chains. There are no reported observations of perfectly crystalline chains in gel phases in aqueous systems of complex lipids of the type that occurs in membranes. The fact that the chains in membranes are highly disordered even when they are described as crystalline has certain functional aspects. The corresponding hexagonal arrangement of extended hydrocarbon chains also exists in monomolecular films at the air–water interface, and, because of its liquidlike properties, it was classified as a liquid phase in the monolayer phase nomenclature by Harkins (*21*). This should be kept in mind when lateral diffusion in membranes is considered. When membrane lipids are in a so-called crystalline state, the hexagonal chain arrangement allows a considerable degree of lateral movement of the molecules. Many recent studies indicate that a certain portion of the hydrocarbon chains in membranes are in a crystalline state (*cf.* Ref. *10*). The segregation of chains into ordered and disordered regions and the dynamic properties involved in these transitions are therefore important in understanding the structure and functions of biomembranes.

Cholesterol has a profound effect on the transition between ordered and disordered chains. Calorimetric measurements of aqueous lecithin–cholesterol phases reveal a broadening in the transition temperature range and a reduction in transition energy with increasing cholesterol content; at a 1:1 molecular ratio, no transition is observed (*22*). Raman spectroscopy of the same system demonstrates that the change from gauche to trans conformation of the chains at the chain crystallization temperature of lecithin bilayers is successively lost when cholesterol is added; this can be interpreted as the formation of a glassy state when cholesterol is added (*23*). It seems obvious that the rigid cholesterol skeleton must reduce the mobility of the hydrocarbon chains in the liquid state, and also that the cholesterol molecules cannot be accommodated into any of the known close-packing arrangements of hydrocarbon chains. That molecular separation into cholesterol and phospholipid regions is not occurring at the chain crystallization temperature must result from a strong association of the molecules of the polar head groups. One function of cholesterol in membranes might be to affect the transition crystalline $\rightleftarrows$ melted chains and the segregation of the lipid molecules into ordered and disordered regions.

In a recent x-ray diffraction study of oriented multilayers of a lecithin analog (1-oleoyl-2-*n*-hexadecyl-2-deoxyglycero-3-phosphorylcholine) by Lesslauer *et al.* (*25*), the packing of the hydrocarbon chains was discussed on the basis of the observed high-angle diffraction. The available information on hydrocarbon chain packing in lipids (*cf.* Refs. *26, 27, 28*), however, was not taken into consideration. Since the order in phospholipid bilayers is a problem of general significance in connection with membranes, we discuss the structure of the hydrocarbon region in this particular case.

The observed short spacings (at 0% relative humidity) at 4.14 and 4.64 A attributable to chain distances in the plane perpendicular to the chains were interpreted as corresponding to an orthorhombic subcell with lateral axes of 8.26 and 5.59 A. It can be assumed that such a cell is not possible in the case of crystalline chains since the cross-sectional area per chain is about 23 A^2. Even in the loose hexagonal chain arrangement, where the chains have rotational freedom, the cross-sectional area per chain is only about 19 A^2. Seven different modes of packing the chains were observed; these were designated $T\|$, $M\|$, $O\perp$, $O'\perp$, $O\|$, $O'\|$, and H in accordance with subcell symmetry and chain plane direction (*29*). The different types of chain packing can be identified by their x-ray short-spacing diffractions although the related subcells $O\perp$ and $O'\perp$ or $O\|$ and $O'\|$ cannot be separated unambiguously (*27*). The two dominant short spacing lines at 4.64 and 4.14 A do not agree with those of the known chain-packing subcells, but this indicates that there are two different types of chain arrangement. The most common chain-packing subcell for sat-

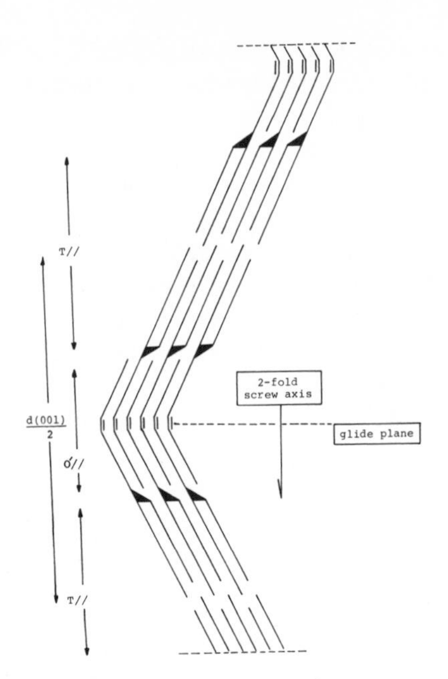

Figure 15. Hydrocarbon chain structure of 2-oleyl-distearin based on x-ray diffraction. The short spacing data indicate that the two types of chain packing (T_{II} and O'_{II}) are attributable to separation of the chains into bilayers with saturated chains and monolayers with unsaturated chains. The long spacing data indicate that the unit layer consists of three chain layers.

urated normal chains is the triclinic $T_{||}$ which has one dominant short-spacing line at about 4.6 A. $T_{||}$ and the orthorhombic packing $O_{\perp}$ give the closest chain packing. Introduction of irregularities in chain structure as well as impurities in the chain layers result in a larger cross-sectional area per chain, and the hexagonal arrangement (with one short spacing line at about 4.15 A) is then favored.

With lipid molecules that contain both saturated and unsaturated chains, a chain-sorting mechanism often results in separation of saturated and unsaturated chains in different layers. No complete crystal structure of such a complex lipid is known. On the basis of x-ray powder data for 2-oleyl-distearin, it is possible to derive the principal molecular arrangement (Figure 15) (*28*), and the separation of saturated and unsaturated chains can be seen. With regard to chain packing, there are thus certain advantages to an extended molecular conformation of crystalline membrane phospholipids that have one saturated and one unsaturated hydrocarbon chain. The possibility of a lipid phase transition between the normal bilayer conformation in which the polar groups form the outer surfaces and an extended form in which the polar groups are in the middle has some support in the monolayer behavior of diglycerides (*30*). Shielding of the polar head groups in a membrane region with lipid bilayers and liquid chains so that their environment becomes less polar might result in a phase transition into the extended form with crystalline chains. Repeated transitions of this type are equivalent to flip-flop movement of a certain portion of the lipid molecules, and transport of adsorbed ions or small molecules might be achieved in this way. There is recent experimental evidence that transport of ions across lipid bilayers can be effected by phospholipid molecules (*31*).

Electrical Properties

The electrical properties of black lipid membranes (BLM's) have probably been studied more than those of other lipid systems because of the great similarity between BLM's and cell membranes. The electrical properties of BLM's were reviewed extensively by other authors (*32, 33, 34, 35*), and we shall therefore describe the electrical and physical properties of lipids which are not generally touched upon in connection with BLM's. We also concentrate on those properties which are intimately related to the different states of order in lipid systems.

Electrical Properties of Lipid–Water Systems. The electrical properties of different mesophases were recently reviewed by Winsor (*36*). Measurements demonstrated for example (a) that phase transitions are seen as changes in electrical conductivity and (b) that the conductivity in amphiphilic mesophases may show a large anisotropy depending on the structure of the mesophase. We made some measurements on lamellar aerosol–OT (A–OT) water systems in order to obtain information about the conductance of the rather thin water layers in these systems (*37*). Oriented samples were prepared by placing a small amount of A–OT between two glass slides and using thin gold wires (50 μm) as spacers and electrodes. In order to avoid polarization effects, conductivity was measured by putting a square current through the sample and displaying the voltage drop across a series resistor on an oscilloscope. The voltage drop at $t = 0$ (which corresponds to infinite frequencies) contains the information about the true conductance of the sample. The data from these measurements are presented in Figure 16; the conductivity is plotted *vs.* $1/d$ where d is the lamellar repeat distance (*38*). Arrhenius plots have a breaking point at a certain temperature for some of the samples. It was concluded in Ref. *37* that the conductance of the water lamellae cannot be explained by normal sodium conductance and that the narrow water channels affect either the mobility of sodium ions and/or the number of free sodium ions. Other conduction mechanisms (*e.g.* proton jumping along an ordered water structure) are also possibile.

Some measurements were also made on lamellar tetradecyl amine–water systems (*see* section on The Gel State in Lipid–Water Systems) with 88–95% water using the method described in Ref. *37*. In this case, the water layer was about 1500–1600 A thick, and no counterions were present. Conductivity in the lamellar region is therefore expected to be caused by protons jumping between the amine groups of the lipid. Figure

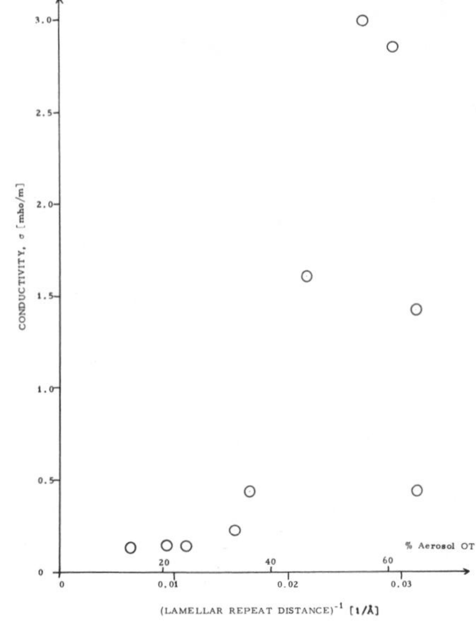

Figure 16. Plot of the conductivity of aerosol–OT (A–OT) samples vs. 1/d *where* d *is the lamellar repeat distance (38). Samples with 65.1% A–OT were sometimes bistable, switching between high and low conductivity.*

Journal of Colloid and Interface Science

17, an Arrhenius plot of the natural logarithm of the conductivity *vs.* $1/T$ for two samples, indicates what might happen at a phase transition. The conductivity changes as well as the slope of the $1/T$ plot. Furthermore, it was observed that when the temperature was increased, the conductivity first decreased rapidly at the phase transition and then increased to the value indicated in the figure. Conductivity behavior was similar for all samples tested. Room temperature conductivity was also about the same (0.005–0.008 1/Ωm); activation energy of conductivity was about 0.09–0.12 eV below the phase transition and in the order of 0.2–0.4 eV above the phase transition.

If we assume that the conductivity before the phase transition is attributable to protons jumping along the lipid layers and that the number of protons per lipid layer is given by the number of ionized amine groups (*see* above), then the mobility of the protons is about one-tenth that of protons in water and the activation energy is similar to that of protons in water (in the same temperature region). Another possibility is that only about 10% of the ionized amine groups act as proton donors. If this interpretation is correct, then protons may jump along the lipid–water interface. Comparing these findings with those for A–OT, we find that water structure becomes important in determining conductivity when the thickness of the water layer is small.

Electrical Properties of Lipid Multilayers. By the Langmuir–Blodgett technique (*39, 40*), it is possible to obtain well ordered multilayers of lipids on solid supports like glass, metals, and semiconductors. A number of physical investigations of these types of films have been made, some of which are reviewed briefly here. Blodgett (*41*) used this technique to prepare organic anti-reflection coatings on glass slides. Many of the physical investigations of fatty acid multilayers were reviewed by Kuhn and Möbius (*42*). Kuhn and co-workers in recent years used the Langmuir–Blodgett technique in very ingenious ways (they incorporated layers of dye molecules in fatty acid multilayers) to study, among other things, how excitation energy is transferred between molecules (*42, 43*). Mann *et al.* (*44*) investigated the tunneling of electrons through fatty acid monolayers; they found that the experimental data agreed with the exponential decrease in conductance *vs.* thickness that is predicted by tunnel theory.

A number of other investigations of the electrical properties of lipid mono- and multilayers were published recently. It is obvious from studies of the conductivity of thin Langmuir films that the electrical properties of metal–organic layer–metal structures can be described by well known concepts from solid state physics, like Schottky injection of electrons from the metal into the lipid film (*45, 46, 47*). Measurements of dielectric losses in calcium stearate and behenate indicate the presence of movements of dipoles in the organic molecules, and loss peaks connected with the amorphous and crystalline parts of the layers were identified (*48*).

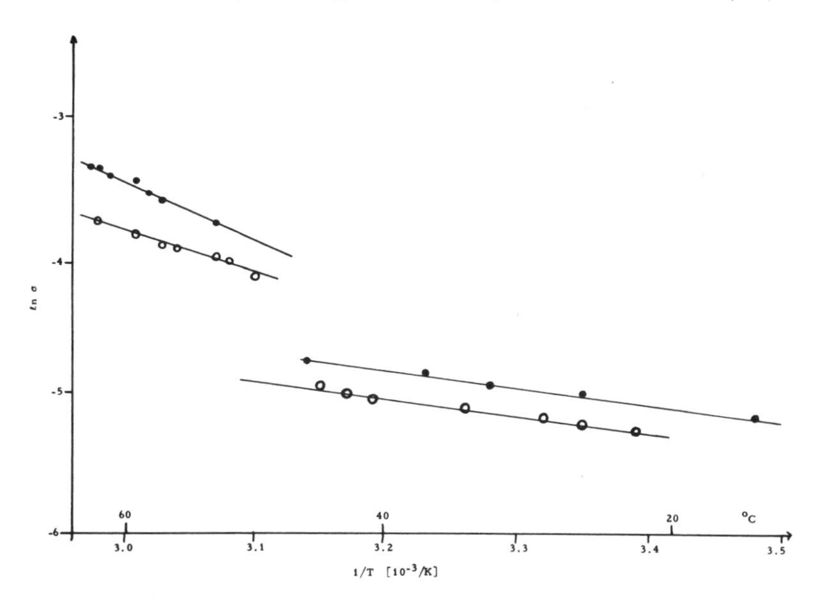

Figure 17. Arrhenius plot of conductivity 1/T for a tetradecylamine–water system. Conductivity was measured during a slow increase in temperature. ●: 92% water; and ○: 88% water.

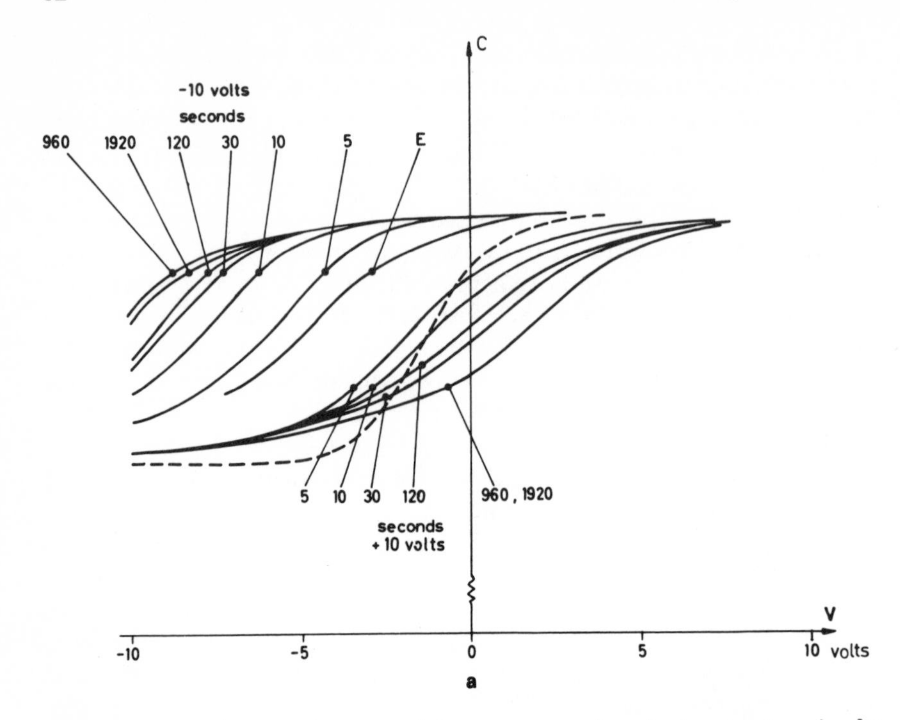

Figure 18. C(V) curves for a metal–Ba stearate semiconductor structure (multilayer thickness, ~1000 A). Capacitance levels are indicated; the max/min ratio depends on the parameters of the structure, and the absolute values of the capacitance depend of course on the area of the metal contact (a mercury probe). Different areas of the same sample were used to obtain the curves in the top and bottom figures. ---: an ideal, theoretical C(V) curve.

a: curve E: the experimental equilibrium C(V) curve that was obtained when the structure was kept for a long time without voltage being applied. The curves displaced to the left of Curve E were obtained by applying negative voltage (10 V) pulses (metal negative) across the structure for the indicated periods of time. The curves displaced to right of Curve E were obtained by applying positive voltage (10 V) pulses. These curves indicate that, during negative voltage pulses, holes are injected into the fatty acid whereas electrons are injected during positive voltage pulses; the shift along the voltage axis from Curve E is a direct measure of the amount of charge injected.

Procarione and Kaufmann (*49*) studied the electrical properties of phospholipid bilayers between metal contacts. They observed, for example, irregularities in current and capacitance *vs.* temperature data which may be the result of phase transitions in the lipid bilayer. They also observed that both temperature-independent (tunneling) and temperature-dependent conduction processes with an activation energy of 0.65 eV were important.

Recently the so-called C(V) technique was introduced to the study of lipids (*50, 51, 52*). This is a method borrowed from semiconductor

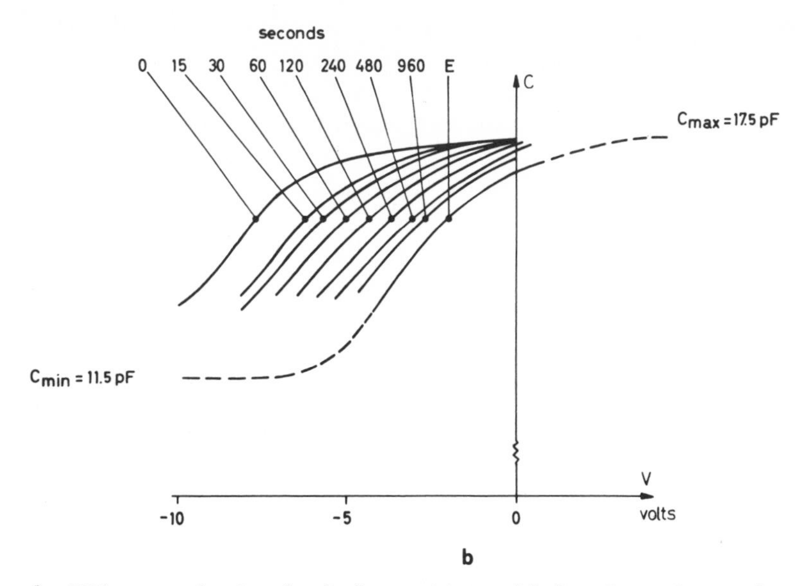

b: C(V) curves showing the discharge of injected holes. Curve O was obtained by applying −10 V for 2 min. The other curves were obtained at the indicated times after the voltage had been removed.

surface studies, and it enables one to study the amount of charge stored or introduced in an insulator on top of a semiconductor. The technique is described in Refs. *50* and *52*. In principle, the capacitance of a metal–organic layer semiconductor structure is measured as a function of the potential difference between the metal and the semiconductor. A theoretical ideal curve is represented by the dashed line in Figure 18a. For an *n*-type semiconductor, a positive potential on the metal electrode attracts electrons to the semiconductor surface, and we measure the capacitance of the insulator itself. If the potential becomes negative, a region depleted of electrons is created at the semiconductor surface; we measure the insulator capacitance in series with the capacitance of the depleted region, and the total capacitance decreases. At large negative potentials, the width of the depleted region becomes constant, independent of voltage, and we measure a constant (small) capacitance. If there are charges in the insulator, however, the capacitance curve is displaced in voltage.

It is easy to understand that positive charges in the insulator mean that a larger negative potential can be applied on the metal before electrons are repelled from the semiconductor surface, and therefore the C(V) curve is displaced to the left. Obviously, negative charges in the insulator displace the curve to the right. By making metal–stearate (behenate, etc.) semiconductor structures (with an insulator thickness of

500–100 A) by the Langmuir–Blodgett technique, we were able to demonstrate that electrons and holes can be introduced into fatty acid multilayers by applying large voltage pulses across the structure (*50, 52*). The curves in Figure 18a were obtained from such an experiment: negative voltage pulses introduced positive charges (holes) in the fatty acid multilayers and positive voltage pulses introduced negative charges (electrons) in the multilayers.

Measurements at different temperatures revealed that the injection process was temperature-dependent with an activation energy of about 0.65 eV (*52*). This activation energy is consistent with conductivity measurements on metal–fatty acid–metal structures (*46, 47*) and also with the activation energy of the (high temperature) conductance of phospholipid bilayers (*49*). The activation energy may be the difference between the semiconductor Fermi level and the valence (and conduction) band edges in the fatty acid, or it may be the energy difference between some localized electronic states in the forbidden band gap of the fatty acid and the band edges.

We also studied (*50, 52*) how long the introduced charges stayed in the fatty acid. It was concluded that memory times in the order of tenths of seconds could be obtained at room temperature (Figure 18b). No significant difference between fatty acids of different hydrocarbon chain lengths was observed in our experiments. The experimental results compare favorably with simple theoretical models similar to those applied to a new type of semiconductor memory device (*53*).

More interesting, however, was the fact that minor changes in composition and structure of the films changed charge storage properties markedly. Introduction of a small amount of erucic acid, an unsaturated fatty acid, into a behenic acid multilayer changed the activation energy to 0.6 eV which changed significantly the speed of charge injection. Lipids with fluorine-substituted hydrogen of the methyl group could not store charge to the same extent as the pure lipids. Furthermore, lipid-soluble gases like ether, iodine, and halothane changed the charge storage properties markedly; at sufficiently large concentrations, charge storage ability almost disappeared. Water vapor, on the other hand, seemed only to increase somewhat the speed of charge injection.

These findings suggest that the charge storage ability of fatty acid multilayers depends on the methyl end groups of the hydrocarbon chains, that is, on the methyl gap between two fatty acid layers. This also suggests that conductivity of crystalline fatty acids probably should be greater in a direction perpendicular to the hydrocarbon chains than in a direction parallel to the chains, the conducting charges being trapped at the methyl gap. Although it is rather easy to prepare thin crystalline layers for measurements along the hydrocarbon chains, it is more difficult to prepare and contact layers for perpendicular measurements. In some preliminary studies (*54*), we made thin layers between glass slides by melting small amounts of fatty acids between them. By slowly heating the sample from one end to the other many times (zone refining), a rather good crystalline layer was obtained with the molecules oriented perpendicular to the glass slides. Gold wires were used as spacers and contacts for measurements perpendicular to the molecules, evaporated gold stripes on the glass slides for measurements along the molecules. The experiments demonstrated that the conductivity perpendicular to the hydrocarbon chains was about 20–100 times greater than that along the chains below the melting point and that the two conductivities became equal (the perpendicular decreased as the parallel increased) at the melting point of the stearic acid. The activation energy of the conductivity was about the same parallel and perpendicular to the chains (0.9–1 eV), and it did not change appreciably on melting. The conductivity along the (unmolten) chains was of the same order as that of Langmuir films ($\sim 10^{-12}$ $1/\Omega m$ at room temperature). Even if the prepared samples were far from perfect, the findings certainly indicate that the conductivity perpendicular to the hydrocarbon chains is greater than that parallel to the chains.

In the perpendicular direction, there may be conduction along the polar end group or by change transfer between the hydrocarbon chains. Proton conductivity is also a possibility. No gas evaluation and no increase in resistance from polarization effects were noted, however, in our preliminary experiments. Recently we have started some studies on electrolyte–lipid semiconductor structures. We observed, for example, interesting differences in the effect of divalent ions on different lipid monolayers (*55*).

We believe that the type of investigations that are outlined briefly above provide interesting new information on the properties of lipids from both technical and biophysical points of view. Measurements on electrolyte–lipid semiconductor systems should provide useful information complementary to that obtained from BLM investigations. Furthermore, the gas sensitivity of the electrical properties of lipids could be utilized in practical devices.

Biological Implications of Structural and Electrical Properties of Lipids. It is rather obvious that the structure of lipids is very important in connection with the function of living cells since most physiological processes occur in lipid environment. There is, for example, evidence that lipid–protein complexes are necessary for the proper functioning of mitochondria (*56*). Although lipids are most important in providing a suitable material for functional complexes (ionic channels, electron transport systems, receptor units, etc.), their own physical properties are certainly not unrelated to the function of living systems. Furthermore, different lipid systems may serve as models from which extrapolations to a completely different situation can be made. The thin water channels in certain lamellar systems may be similar to the ionic channels of nerve membranes where the channels may actually be created by protein molecules. Proton conductivity along lipid–water interfaces, interesting as such, may also model the proton conductivity of certain proteins or polypeptides with proper side groups. The electronic properties of lipids may be interesting in connection with electron donors and electron acceptors on opposite sides of a membrane. Is it, for example, possible to transport electrons across a lipid, thereby creating a difference in potential? Interesting in this connection is the fact that electrons may tunnel against a potential gradient as long as suitable energy levels are present. Are the electronic traps that are present in a membrane caused by the lipids or by macromolecules with properties similar to those of the lipids with regard to charge storage? And, if so, is it not conceivable that many of the triggering functions of cell membranes (nerve walls, nerve endings, retinal membrane, smell receptor sites, etc.) have an electronic origin? It is not possible that charge storage is a short term memory used in some of the processes in our brain? These are just a few of the questions which are inspired by the electrical properties of lipids and lipid–protein complexes. The first assumption of electronic processes in living systems was probably that of Szent-Györgyi about semiconduction in proteins (*57*), and he and others demonstrated afterwards that proteins may be a part of charge transfer complexes and that their conductivity increases when they are treated suitably (doped) (*58*). Rosenberg *et al.* discussed semiconduction in proteins in some detail (*60*). The observed photosensitivity of BLM's that contain chlorophyll (*59, 61, 62*) indicates that electronic processes occur across the membrane. The evidence for electronic processes in living systems is thus very great, and of course electronic processes are known to take place in the chloroplasts and in the mitochondria.

We recently proposed a completely electronic model for the excitability of nerve membranes that is based on the assumption of electron-donating, electron-accepting, and electron-storing properties of macromolecules or of protein–lipid complexes which constitute the ionic channels of the nerve membrane (*63*). This model, which is based on simple physical concepts with easily defined parameters, reproduces the empirical Hodkgin–Huxley equations rather well and also explains how different types of drugs may work on nerves. The model is easily extended to other excitable complexes like the receptor protein complex at nerve synapses and the rodopsin molecules in the retina. Nor is it inconceivable to build a model for the function of smell that is based on electronic triggering of ionic channels which are affected by molecules adsorbed onto or dissolved in the lipids of all membranes. This rather speculative section is only to remind us that, even if most of the living processes do occur in a watery environment with ions serving as important charge carriers, the electronic properties of macromolecules and lipid bilayers may be important in ways not normally thought of. It is needless to say that the electronic properties are intimately related to the structural properties of the macromolecules, the lipids, and their complexes.

We shall end this section by showing how very general physical phenomena may explain an experimental observation on nerve membranes which, to our knowledge, was not explained satisfactorily before (*64*). When we study the current through a nerve membrane at a potential difference so that the ionic channels are open, we find a noise in the current whose power spectrum is proportional to $1/f$ where f is the frequency. This $1/f$ noise, which can be observed down to very low frequencies ($\sim$ 10 Hz), is rather puzzling since it is known that (a) the gating processes (opening of ionic channels) have time constants in the order of 0.1–1 msec, and (b) the ions flowing through an open channel spend about 0.1 msec in the channel. Hence, these conditions do not produce the large time constants that are necessary in order to explain $1/f$ noise down to 10 Hz (time constants about 0.1–1 sec).

The $1/f$ noise is often attributed to the potassium current through the nerve membrane, and it is then related to the number of ions that pass a potassium channel per unit time, *i.e.* the conductance of the (open) channel. It was recently demonstrated that $1/f$ noise can be explained by simple assumptions about the properties of the lipids and the lipid–channel molecule complexes in nerve membranes (*65*, *66*). First of all, we assume that the lipid bilayer or the hydrocarbon chains within it have a liquid crystalline property in that the free energy of the bilayer depends on the gradient of the direction of the molecules or segments of the molecules. Secondly, the movement of the chains (or segments of them) is described by a diffusionlike equation that is used to describe fluctuations in liquid crystals. Finally, we assume that the fluctuations in the direction of the lipid molecules (or segments of them) close to an ionic channel modulate the conductivity of the channel. Figure 19 is a very simple diagram of this model. The movements of the lipid molecules may, for example, affect the channel geometrically, or they may move charged groups, thereby changing the electrostatic potential seen by ions moving through the channel. By using these rather simple assumptions, it was possible to determine $1/f$ noise down to low frequencies. The calculation of the noise spectrum is found in Refs. *65* and *67*. Our physical model is an alternative to models based on channel statistics; it provides a basis for further studies of $1/f$ noise in nerve membranes. Although it has all the general features

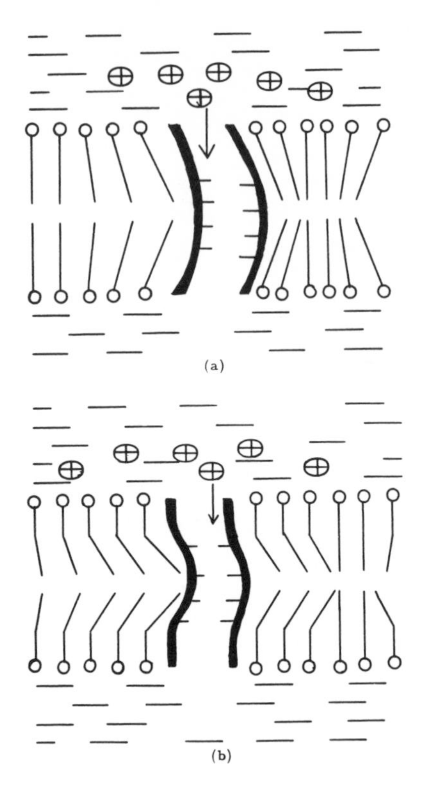

Figure 19. Schematic of how an ionic channel in a nerve membrane can be affected by the direction of the lipid molecules (a) or of their segments (b). The direction of the molecules can determine the geometrical configuration of the channel and/or change the position of the charged groups that affect the potential seen by a positive ion, thereby affecting the mobility of a channel.

of the experimentally observed noise, there are several possibilities which were not accounted for. For example, do the ionic channels (and the current through them) affect the movements of the lipid molecules? In our simple model, the noise power is proportional to I_K^2 where I_K is the potassium current. If the potassium current consists of the sum of two independent currents, one outward and one inward (because of active transport), do we get a $1/f$ noise even when $I_K = 0$? The tentative answer is yes, but this remains one of the interesting questions which should be answered by future studies.

An interesting experiment in connection with $1/f$ noise in nerve membranes would be to study the noise in current through BLM's that contain an ionophore like gramicidin. It is known that the conductivity of BLM's containing gramicidin does not change appreciably at the melting point of the hydrocarbon chains of the BLM, which indicates that gramicidin acts as an ionic channel (68). If the movement of the hydrocarbon chains modulates the conductivity of the gramicidin ionic channel, we would expect, however, the magnitude of the noise to increase at the melting of the hydrocarbon chains.

Final Remarks

We have summarized some of the present knowledge of the structural and physical properties of liquid crystalline systems. We are aware that a number of interesting findings and methods of investigations were not mentioned. The main purpose of this communication however, was to discuss briefly our own and closely related fiindings and ideas. We have also speculated about the possible biological significance of properties of lipids and lipid–water systems. It is evident that model systems and physical properties of lipids that are not generally thought of may be of interest in connection with the modelling and the understanding of cell membranes.

Literature Cited

1. Chapman, D., *Proc. Chem. Soc. London* (1958) 784.
2. Luzzati, V., Mustacchi, Skoulios, A., Husson, F., *Acta Crystallogr.* (1960) **13**, 660.
3. Small, D., *J. Lipid Res.* (1967) **8**, 551.
4. Luzzati, V., Gulik-Krzywicki, T., Tardieu, A., *Nature London* (1968) **218**, 1031.
5. Chapman, D., Williams, R. M., Ladbrooke, B. D., *Chem. Phys. Lipids* (1967) **1**, 445.
6. Luzzati, V., Tardieu, A., Gulik-Krzywicki, T., *Nature London* (1968) **217**, 1028.
7. Larsson, K., *Z. Phys. Chem.* (1967) **56**, 173.
8. Larsson, K., *Chem. Phys. Lipids* (1972) **9**, 181.
9. Lundberg, B. *Chem. Phys. Lipids* (1973) **11**, 219.
10. Small, D. M., "Surface Chemistry of Biological Systems," p. 55, Plenum, New York, 1970.
11. Small, D. M., Shipley, G., *Science* (1974) **185**, 222.
12. Small, D. M., Bourges, M., Dervichian, D. G., *Nature London* (1966) **211**, 816.
13. Rand, R. P., *Biochim. Biophys. Acta* (1971) **241**, 823.
14. Rand, R. P., SenGupta, S., *Biochemistry* (1972) **11**, 945.
15. Tanford, C., *J. Mol. Biol.* (1972) **67**, 59.
16. Larrsson, K., *Chem. Phys. Lipids* (1973) **10**, 165.
17. Larsson, K., Rand, R. P., *Biochim. Biophys. Acta* (1973) **326**, 245.
18. Larsson, K., *Chem. Phys. Lipids* (1974) **12**, 176.
19. Abrahamsson, S., Larsson, K., Pascher, I., unpublished data.
20. Larsson, K., Krog, N., *Chem. Phys. Lipids* (1973) **10**, 177.
21. Harkins, W. D., "The Physical Chemistry of Films," 2nd ed., Reinhold, New York, 1954.
22. Ladbrooke, B. D., Williams, R. M., Chapman, D., *Biochim. Biophys. Acta* (1968) **150**, 333.
23. Lippert, J. L., Peticolas, W. L., *Biochim. Biophys. Acta* (1972) **282**, 8.
24. Oldfield, E., *Science* (1973) **180**, 982.
25. Lesslauer, W., Slotboom, A. J., de Haas, G. H., *Chem. Phys. Lipids* (1973) **11**, 181.
26. Chapman, D., "The Structure of Lipids," Methuen, London, 1964.
27. Abrahamsson, S., Ställberg-Stenhagen, S., Stenhagen, E., "The Higher Saturated Branched-Chain Fatty Acids," *Prog. Chem. Fats Other Lipids* (1963) **7**.

28. Larsson, K., *Fette Stefen Anstrichm.* (1972) **74,** 136.
29. Larsson, K., *Acta Chem. Scand.* (1966) **20,** 2255.
30. Larsson, K., *Biochim. Biophys. Acta* (1973) **318,** 1.
31. Harris, R. A., Farmer, B., *Lipids* (1974) **9,** 717.
32. Läuger, P., *Naturwissenschaften* (1970) **57,** 474.
33. Läuger, P., *Science* (1972) **178,** 24.
34. Jain, M. K., "The Bimolecular Lipid Membrane," van Nostrand–Reinhold, New York, 1972.
35. "Membranes," vol. 2, "Lipid Bilayers and Antibiotics," G. Eisenman, Ed., Marcel Dekker, New York, 1973.
36. Winsor, P. A., in "Liquid Crystals and Plastics Crystals," G. W. Gray and P. A. Winsor, Eds., vol. II, p. 122, Wiley, London, 1974.
37. Lundström, I., Fontell, K., Lexell, L., Mattiasson, R., unpublished data.
38. Fontell, K., *J. Colloid Interface Sci.* (1973) **44,** 318.
39. Blodgett, K., *J. Am. Chem. Soc.* (1935) **57,** 1007.
40. Langmuir, I., Schaefer, V. J., *J. Am. Chem. Soc.* (1937) **59,** 1406.
41. Blodgett, K., *Phys. Rev.* (1939) **55,** 391.
42. Kuhn, A., Möbius, D., *Angew. Chem.* (1971) **83,** 672.
43. Bücher, H., Drexhage, K. H., Fleck, M., Kuhn, H., Möbius, D., Schäfer, F. P., Sondermann, J., Sperling, W., Tillmann, P., Wiegand, J., *Mol. Cryst.* (1967) **2,** 199.
44. Mann, B., Kuhn, H., Szentpaly, L. V., *Chem. Phys. Lett.* (1971) **8,** 82.
45. Handy, R. M., Scala, L. C., *J. Electronchem. Soc.* (1966) **113,** 109.
46. Horiuchi, S., Yamaguchi, J., Naito, K., *J. Electrochem. Soc* (1968) **115,** 634.
47. Wei, L. Y., Woo, B. Y., *Biophys. J.* (1973) **13,** 877.
48. Marc, G., Messier, J., *J. Appl. Phys.* (1974) **45,** 2832.
49. Procarione, W. L., Kaufmann, J. W., *Chem. Phys. Lipids* (1974) **12,** 251.
50. Lundström, I., McQueen, D., *Chem. Phys. Lipids* (1973) **10,** 181.
51. Tanguy, J., *Thin Solid Films* (1972) **13,** 33.
52. Lundström, I., Stenberg, M., *Chem. Phys. Lipids* (1974) **12,** 287.
53. Lundström, I., Svensson, C., *IEEE Trans. Electron Devices* (1972) **19,** 826.
54. Stenberg, M., Lundström, I., unpublished data.
55. Stenberg, M., Lundström, I., Sheorey, U., unpublished data.
56. Jost, P. C., Capaldi, R. A., Vanderkovi, G., Griffith, O. H., *J. Supramol. Struct.* (1973) **1,** 269.
57. Szent-Györgyi, A., *Nature London* (1941) **148,** 157.
58. Szent-Györgyi, A., McLaughlin, J., *Proc. Natl. Acad. Sci. U.S.A.* (1972) **69,** 3510.
59. Davies, K. M. C., Eley, D., Snart, R. S., *Nature London* (1960) **188,** 724.
60. Rosenberg, B., Postow, E., *Ann. N. Y. Acad. Sci.* (1969) **158,** 161.
61. Tien, H. T., Venna, S. P., *Nature London* (1970) **227,** 1232.
62. Ilani, A., Berns, D. S., *Biophysik* (1973) **9,** 209.
63. Lundström, I., Stenberg, M., unpublished data.
64. Bird, J. F., *Biophys. J.* (1974) **14,** 563.
65. Lundström, I., McQueen, D., *J. Theor. Biol.* (1974) **45,** 405.
66. Lundström, I., McQueen, D., *Intern. Liquid Crystal Conf. 5th, Stockholm, 1974,* abstract p. 185.
67. Lundström, I., McQueen, D., Klason, C., *Solid State Communi.* (1973) **13,** 1941.
68. Krasne, S., Eiseman, G., Szabo, G., *Science* (1971) **174,** 412.

RECEIVED November 19, 1974.

5

Recognition of Defects in Water–Lecithin Lα Phases

M. KLÉMAN and C. COLLIEX

Laboratoire de Physique des Solides, Université de Paris-Sud,
91405-Orsay, France

M. VEYSSIÉ

Laboratoire de Physique de la Matière Condensée, Collège de France,
75231 Paris Cedex 05, France

The textures in homeotropic lamellar phases of lecithin are studied in lecithin–water phases by polarizing microscopy and in dried phases by electron microscopy. In the former, we observe the Lα phase (the chains are liquid, the polar heads disordered)—the texture displays classical Friedel's oily streaks, which we interpret as clusters of parallel dislocations whose core is split in two disclinations of opposite sign, with a transversal instability of the confocal domain type. In the latter case, the nature of the lamellar phase is less understood. However, the elementary defects (negative staining) are quenched from the Lα phase; they are dislocations or Grandjean terraces, where the same transversal instability can occur. We also observed dislocations with an extended core; these defects seem typical of the phase in the electron microscope.

Textures of lyotropic mesophases have been the object of numerous observations by optical (*1, 2, 3*) and electronic (*4, 5, 6, 7*) microscopy. Except for the pioneering work of Lehmann (*1*) and Friedel (*2*) who intended to identify the various kinds of defects which constitute the textures, the purpose of these observations was to recognize the different existing phases—lamellar, hexagonal (or in the soaps language: neat phase, median phase, etc.)—in correlation with x-ray data.

A similar aim has prompted our interest in the properties of lyotropics in their lamellar modification, *i.e.* in this same structure that is

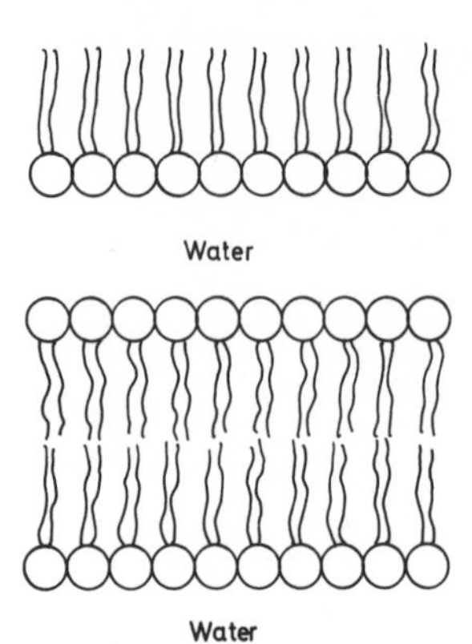

Figure 1. Lα lecithin–water phase. The circles represent the polar heads, the wiggly lines the hydrophobic chains.

characteristic of thermotropic smectics, for which studies of defects are currently under way (*8*). There is certainly a fundamental interest in the study of defects. First, let us note that the word defect is not mere terminology which hides our ignorance of crystal imperfections, but rather that defects can be unambiguously defined as topological entities that are related in a specific way to the symmetries of the perfect medium (*9*). Also, their presence must be taken into account in interpreting many physical measurements since they generally create (or relax) long range stresses in a way that is characteristic of the elastic properties of the medium. Annealing of defects has to be understood as a step toward obtaining perfect samples. Their mere presence strongly affects bulk transport properties; our initial interest in the subject was indeed inspired by discussions with Lange and Gary-Bobo who recently studied (*10*) diffusivity in egg-yolk lecithin–water systems as a function of water content.

In this paper we describe two types of observations on egg-yolk lecithin. We present the results of our study of homeotropically oriented samples of Lα phases by polarizing optical microscopy. This study provides evidence, amid apparently nonsimilar aspects, of the existence of an elementary typical object which we have interpreted as a dislocation. We also studied thin samples of stained lecithin in the high vacuum of the electron microscope. In addition to the defects that are typical of this type of sample, we observed the same elementary object as in Lα lecithin.

Optical Studies

Lα water–lecithin is a lamellar structure in which the polar heads (the phosphatidyl choline group of lecithin) constitute two-dimensional disordered arrays in contact with water, whereas the chains are in the molten state in between water layers in disordered moieties (*see* Figure 1). (For a review of x-ray studies of lecithin–water phases, *see* Ref. *11*.)

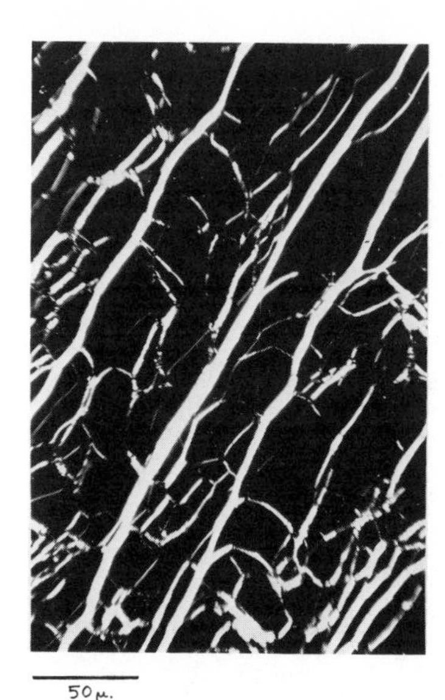

Figure 2. Typical network of defects in a $\Phi_w < 24\%$ sample (crossed polars)

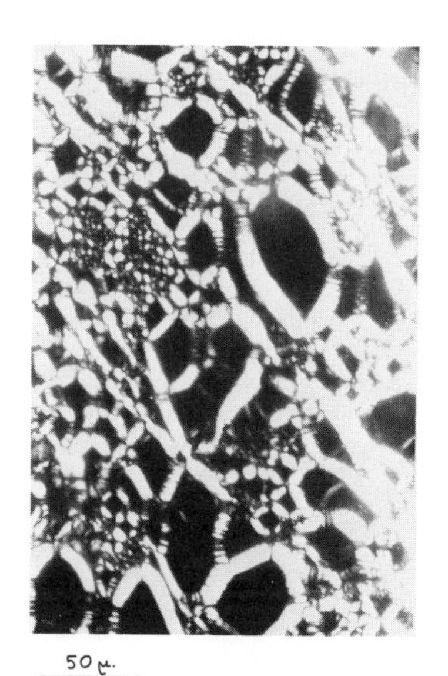

Figure 3. Typical network of defects in a $\Phi_w > 24\%$ sample (crossed polars)

We studied samples of Lα lecithin with 16–29% water at room temperature. The samples were prepared between two microscope glass slides that were first treated with a sulfochromic solution, then carefully rinsed with distilled water and dried. They were observed by transmission polarizing microscopy. Just after deposition between the slides, the samples present a highly disordered texture which strongly diffuses light. After compression (squeezing) and/or shearing parallel to the plates, one obtains an overall homeotropic orientation that is perturbed by defects. The thickness of the samples is in the order of 20μ.

The texture differs significantly with water content. In the less hydrated phases ($\Phi_w < 24$ wt %), the orientation process gives rise to very large homogeneous domains which appear black between crossed polars, with rare bright defect lines organized in a network pattern in the middle of the sample (Figure 2). Such a pattern is mobile when one shears the specimen gently, and it is not attached to the glass plates. At higher water content ($\Phi_w > 24\%$), the same mechanical treatment of squeezing and strong shearing is far less efficient in annealing the original disordered texture. The homeotropic regions are of smaller size, separated by large bunches of defects in a more or less polygonal arrangement (Figure 3). This change in the aspect of the mesophase is rather sharp for the critical concentration of 24%. It is noticeable that the effective shear viscosity of the samples below 24% (as evaluated by hand) is very low, whatever the initial density of defects might be, and

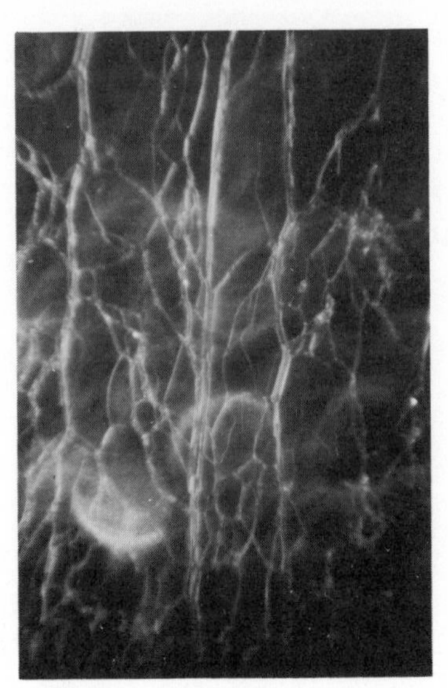

Figure 4. Longitudinal striations in the same field as in Figure 2 (the crossed polars make an angle of 45° with the striations)

it decreases further as the density decreases with subsequent shearing. (Similar annealing of defects by shear was reported in the thermotropic smectic phase of DADB—*see* Ref. *12*). In contrast to this behavior, the initial effective viscosity of high-water-content samples is very large and does not decrease significantly with subsequent shearing; such behavior is related to the observed small decrease in defect density. These facts could be correlated to the anomaly in diffusivity measured by Lange and Gary-Bobo (*10*) at the same critical concentration of 24%, and we then infer that both the diffusion anomaly and the change in visco-elastic properties should be attributed to the same intrinsic reason.

In spite of the striking difference in the aspect of the samples with high and low water contents, the topological nature of the defects forming the texture remains the same; their density and arrangement change only as in a scaling change. In particular, we notice that the following properties of the arrangements are valid for the whole range of water content.

(a) Those lines (*l*) which enter in the formation of the network merge at nodes where their widths add apparently as intensities on a Kirchoff network (conservation law of the widths). This is most visible on well annealed low-water-content samples (*see* Figure 2).

(b) Lines appear to consist of bunches of very thin lines (*see* the longitudinal striations in Figure 4) whose transversal dimensions and distances can be as small as the resolving power of the microscope. These longitudinal striations are generally difficult to observe because of the very low contrast, and special polarization conditions are required.

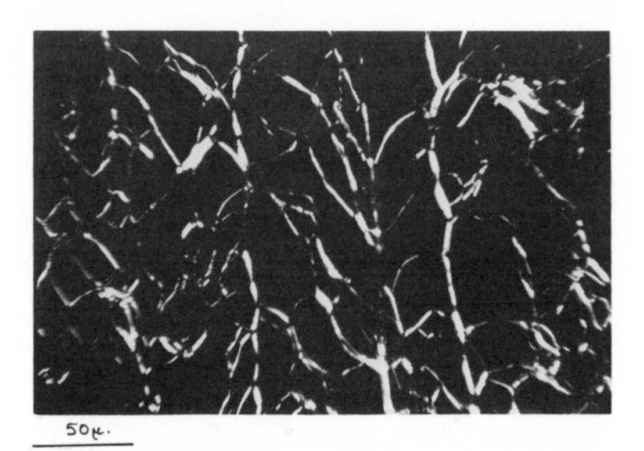

Figure 5. Transversal striations in the same field as in Figure 2 (crossed polars parallel or perpendicular to the striations)

(c) On the same lines (l) a typical transversal structure appears that is most visible when the line is along one of the nicols directions (Figure 5). This transversal structure is clearly reminiscent of the structure described by Lehmann and Friedel as adorning the oily streaks of alcohol–lecithin samples. The relationship between lines (l) and oily streaks is discussed below.

(d) Some confocal domains (c) are clearly apparent, either pinned on lines (l) or (less often) isolated in the bulk (Figure 6). Between crossed polars, they appear as bright circular blobs separated in four equal quadrants by the black brushes of a Maltese cross. This aspect is consistent with the expected shape of confocal domains in an homeotropic specimen. The argument is as follows. Since the asymptotic directions of the hyperbola of c are along the director at a long distance of the

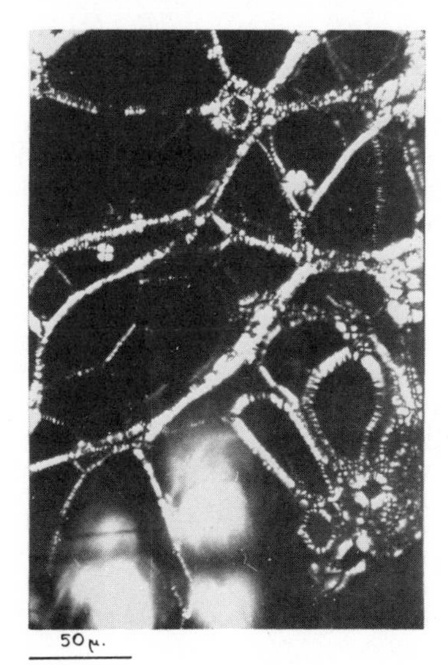

Figure 6. Some visible confocal domains. The transversal striations are clearly visible.

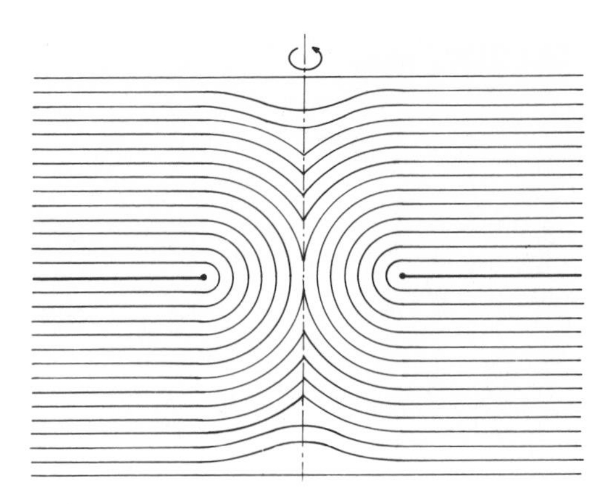

Figure 7. Scheme of a confocal domain in an homeotropic sample. Lamellar details are not featured. Broken line: axis of revolution.

ellipse and since the specimen is homeotropic, the hyperbola is degenerated to a straight line; hence the ellipse is degenerated to a circle, the Dupin cyclides to tori. This is depicted in Figure 7. (For a review of confocal domains geometry, the best introduction is Ref. *2*. For details on the discovery of these topological properties, *see* Ref. *13*. For a more recent view, together with a discussion of the links with dislocations, *see* Ref. *14*.)

(e) The most singular part of a *c* domain is certainly along the axis of revolution. This singularity can be removed and replaced by a less energetic core structure (*see* Figure 8). We believe that such a configuration explains at best the large lines (*L*) which appear mostly in high-water-content samples (Figure 3) as parts of the network, but also on less well annealed parts of low-water-content specimens (Figure 9). Note, however, that a conservation law seems to apply also to *L* lines and that they, as well as *l* lines, have a longitudinal and a transversal texture. We assume that the longitudinal structure in *L* lines is of the type pictured in Figure 8, and we shall argue that, apart from some differences in size especially and in the way of assembling defects, there should not be a fundamental difference between *l* and *L* lines.

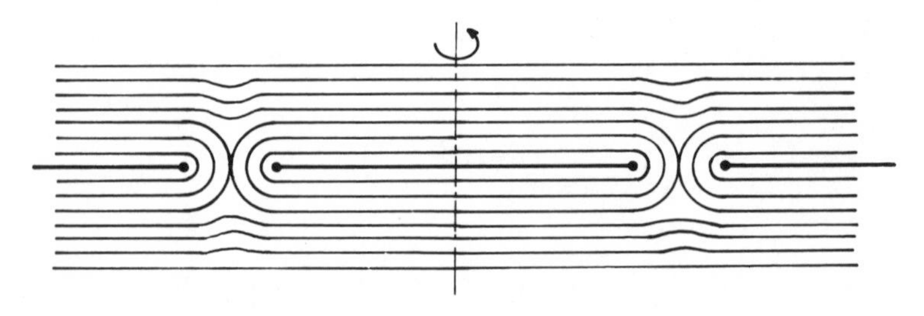

Figure 8. Removal of the core structure of a c *domain. Lamellar are not featured. Broken line: axis of revolution.*

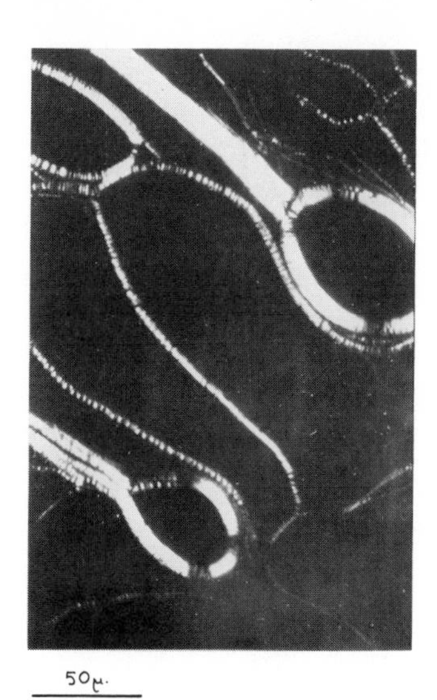

Figure 9. L lines in low-water-content specimen

Discussion of Optical Studies

Friedel (*2*) interpreted the transversal striations on oily streaks as small adjacent confocal domains that have a tendency to gather in lines. We already noted that such a situation exists in DADB (*12*) (but the lines are attached to the surface), and that *c* domains pin up on *l* lines; moreover, oily streaks in cholesterics have clear confocal domains. However, the transversal striations on *l* or *L* lines are not compatible with *c* domains since we do not see there the typical Maltese cross; on the contrary, the hyperbolic directions would be at a small angle to the sample plane if they exist. We do not reject Friedel's explanation, but we must make it compatible with observations, particularly with the longitudinal striations.

According to Figure 8, longitudinal striations are parallel disclination lines. Parallel disclination lines of opposite sign pair in dislocations (Figure 10), and it is more convenient and more physical to consider these latter defects as the elementary objects, since it is well known that the Burgers' vector follows a conservation law of the Kirchkoff type at nodes. (For a review of the idea of disclination, *see* Ref. *15*.) The defect in Figure 8 can itself be divided into two dislocations of opposite sign (Figure 11). Because of the constancy in thickness of the observed samples, each line (*l* or *L*) is a sum of dislocations whose total Burgers' vector is zero or very small, in the order of the boundary surfaces fluctuations. The conservation law observed on widths at nodes is therefore not a necessary condition, but rather it implies that each dislocation keep its individuality at the node.

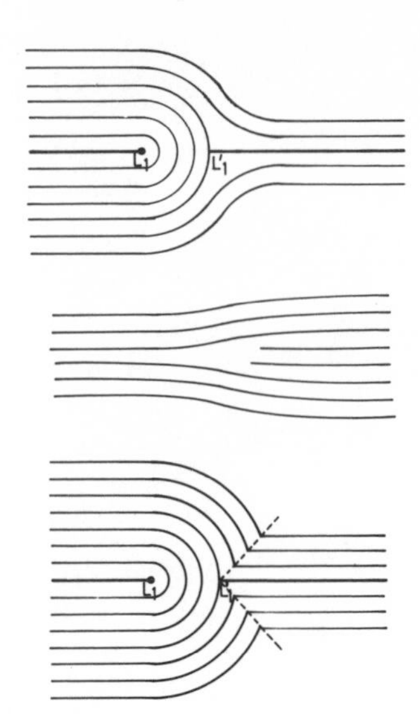

Figure 10. Pairing of two disclination lines of opposite signs (lamellar details are not featured) (top); a less probable model for the core of a dislocation (middle); and focal line appearing on the dislocation in order to release locally deformation energy (bottom)

This last conclusion is certainly in agreement with the core model of the dislocation that we propose in Figure 10 (top). The core model in Figure 10 (middle) is less realistic since it would require that some of the hydrophobic chains be in contact with water; also, such a dislocation would immediately disappear on a similar dislocation of the opposite sign. In contrast, dislocations of the first type (Figure 10, top) keep their individuality easily because their disappearance by fusion with a dislocation of the opposite sign requires some kind of permeation (that is difficult in lyotropics—*see* Ref. *16*) or breaking of the layers.

Concerning the transversal striations, we return to Friedel's basic idea that the easiest deformations possible in a lamellar medium are those which keep the thickness of the layers constant. This is the main property of confocal domains—they involve only splay energy (in the bulk). On the contrary, a dislocation is attended by a deformation in the thickness of the layers which involves an energy of the same order as that of a solid. The model in Figure 10 (top) needs no deformation energy in the region surrounding $L_1(+1/2$ disclination), but the vicinity of L_1' is deformed by a change in layer thickness. We may imagine that such a situation can be relaxed at periodic intervals by focal lines (*see* Figure 10, bottom) (at the expense of line energy and splay energy) which play the role of the hyperbolae, whereas L plays locally the role of the ellipse. The dislocation does not transform entirely to well behaved confocal domains because that would undoubtedly require permeation. However, let us note that in a system in which permeation is easier, the process of transformation can be complete (for cholesterics, *see* Ref. *17;* for thermotropic smectics, *see* Ref. *8*).

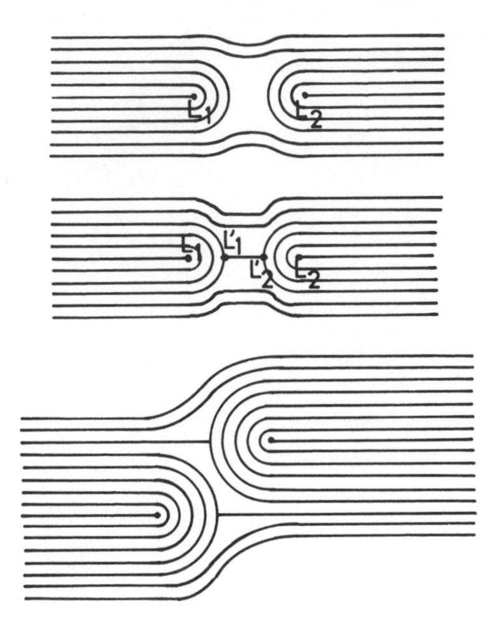

Figure 11. Two disclinations whose topological sum is zero (top); their topological equivalent: two dislocations L_1L_1' and L_2L_2' of opposite sign (middle); and different pairing of two disclinations of opposite sign (bottom)

Electron Microscopy Observations

The specimens are prepared from a colloidal dispersion of egg-yolk lecithin in excess water. In order to obtain reasonable contrast, the water is doped with lead acetate (approximately 0.25 wt %). However, we have no idea of the proportion of lead in the final product, which is prepared as follows. A microscope grid is soaked in the solution and then allowed to dry either in air or on its edge on filter paper. The specimen is then introduced into the high vacuum ($\sim 10^{-6}$ torr) of the electron microscope (Philips EM300) and observed at 100 kV with the help of a tilting stage.

The less well grown parts of the specimens (typical when the specimen dries in air) consist of layered structures perpendicular to the sample surface (Figure 12). Diffraction patterns display only rings at a distance corresponding to the layer thickness; there is no evidence of any kind of order in the layers. This seems to be confirmed by preliminary x-ray diffraction studies of the dried product under a vacuum of $\sim 10^{-4}$ torr—one obtains a layer parameter in the order of 45 A while the chains remain in a molten state (*18*). The aspect shown in Figure 12 does not differ from that obtained by conventional electron microscopic studies by

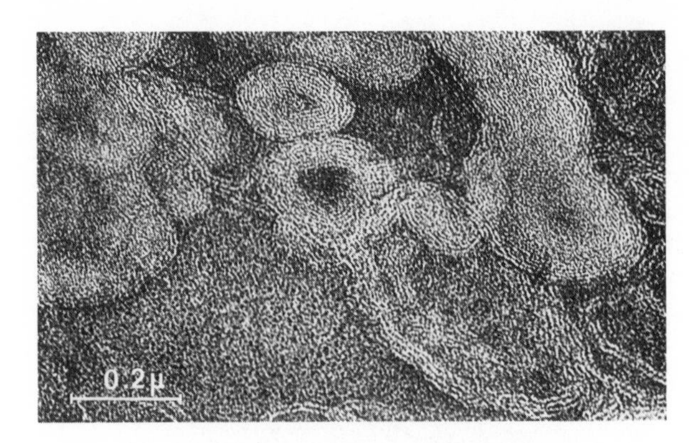

Figure 12. Electron microscopy—the layers are wound up into typical disclinations

freeze–etching or negative staining of lipid–water phases, and it just confirms the tendency of the layers to keep a constant thickness.

More interesting are the specimens obtained by drying on filter paper. In the best parts are beautiful homeotropic areas approximately 2000 A thick which can be observed without any structural change for hours, except for a slight recrystallization of Pb which has no effect on observations. There is no indication whatever of ordering in the layers. We classify the observed defects as (a) Grandjean terraces, (b) dislocation lines, (c) extended disolcation lines, and (d) other objects (*see* Ref. *19*).

Grandjean Terraces. Figure 13 pictures a step on the surface as seen under different conditions of tilt. There is a reversal in contrast from one side of the step to the other, and from positive tilt to negative tilt. The contrast is readily understood if we assume that it is attributable mainly to absorption contrast—the electrons are transmitted better in the regions where the layers are parallel to the beam (channelling effect) than in the region where the planes are perpendicular to the beam. Also, any strong deformation would decrease the contrast in bright field image. We therefore conclude from Figure 13 that the layers of the Grandjean terraces do not stop abruptly, but rather they slope down at an angle of approximately 45°. The thickness of the white or dark bands in the tilted specimen views suggests that the terrace is a few layers thick (2–5 layers). Figure 14 depicts our concept of the distribution of the layers which have to spiral into disclinations in order to explain the contrast. We do not exclude the possibility that symmetrical arrangements keep the thickness constant.

Dislocation Lines. Figure 15 represents a line with characteristic transversal striations. The thickness of the sample changes rapidly from

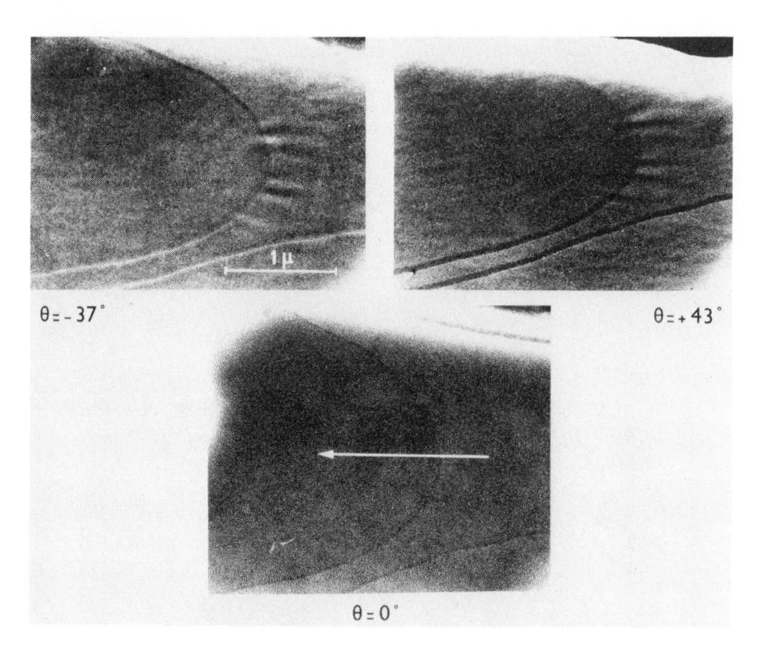

Figure 13. Grandjean terrace (arrow indicates the axis of tilt)

Figure 14. Grandjean terrace—schematic

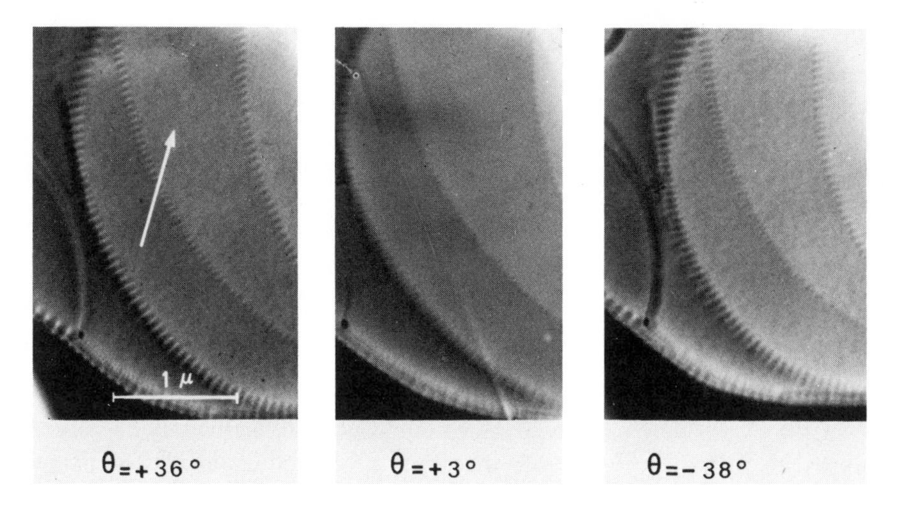

Figure 15. Dislocation lines

Figure 16. Extended dislocations. The thickness of the sample increases from right to left perpendicular to the first family, and from bottom to top past the unique transversal line.

one side of the striations to the other side; hence the Burgers' vector of the defect is large, undoubtedly much larger than that of the Grandjean terrace in Figure 13. It is not possible to ascertain whether the defect in Figure 15 is a dislocation close to the surface (Grandjean terrace) or a dislocation in the bulk. From the comparison of Figures 13 and 15, we conclude that there is a critical size for dislocations above which they display a transversal instability of the type we analyzed above (Figure 10, bottom). The change in contrast of the streaks with different tilt angles suggests some kind of diffraction contrast.

Extended Dislocation Lines. Figure 16 pictures the lines (at tilt angle $\theta = 0$) which are revealed as planar defects in the bulk when the tilting is at other angles. The specimen changes in thickness from edge to edge (thickness contrast is not totally apparent in the figure, but a variation in ribbon width, from one ribbon to another, is visible). Therefore, the planar defects correspond to the location of an extended dislocation with a very peculiar core distribution. On one side of the planar defect there must be, let us say, 20 layers, with 19 layers on the other side; the layers on both sides fit along a common length. This is a kind

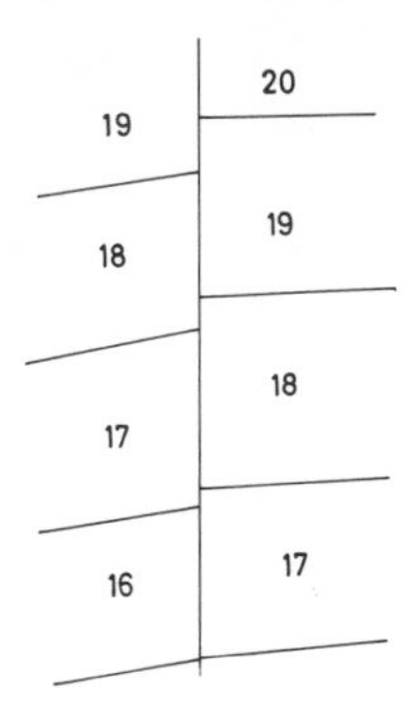

Figure 17. Schematic fitting of regions of different thickness in Figure 16. Horizontal: first family of lines; vertical: unique transversal line.

of epitaxial contact which has the advantage of keeping the layer thickness constant from side to side, while the lamellae suffer a small disorientation of $\sim 1/20$ radian (*see* Figure 18). This disorientation is certainly relaxed at a small distance from the defect. Another planar defect of the same type crosses the first family of planar defects and induces a displacement of the first family.

According to the measured width of the defects, the dislocations to which they correspond have a Burgers' vector of two to five parameters (assuming the parameter equal to 45 A). Figure 17 diagrams the repartition of the lines in the observed region (the numbers correspond to arbitrary values of the thickness measured in parameter units). One notices that the effect of the jogs induced by the unique dislocation is to accommodate thicknesses of the same order of magnitude. It is also reasonable to assert that the jogs are smaller for smaller thickness. The epitaxial fitting is diagrammed in Figure 18. It necessitates contact of the chains with the polar heads, which is possible only if free water is absent.

Conclusion

This study has concentrated on the defects observed in lyotropic lamellar phases, and it has put into evidence the specific character of the textures compared to classical thermotropic smectic phases. In leci-

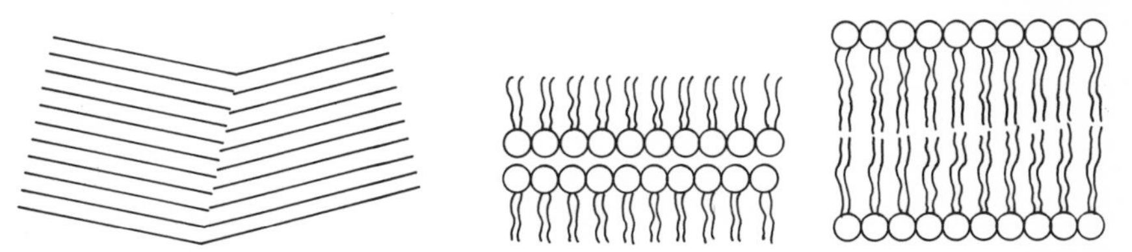

Figure 18. Fitting of the epitaxial layers: planar defect (left) and local configuration of molecules along the planar defect (right)

thin, the essential defect is the dislocation with a core structure that is reasonably interpreted as two disclinations of opposite sign. The transversal instabilities develop with the size increasing with the Burgers' vector; although related to curvature elasticity, they never reach the form of a true confocal domain. The overall texture is a compromise between curvature elasticity and one-dimensional solid elasticity.

Literature Cited

1. Lehmann, O., "Flüssige Kristalle," Wilhelm Engelman, Leipzig, 1904.
2. Friedel, G., *Ann. Phys.* (1922) **18**, 273.
3. Rosevear, F. B., *J. Am. Oil Chem. Soc.* (1954) **31**, 628.
4. Eins, S., *Mol. Cryst. Liq. Cryst.* (1970) **11**, 119.
5. Pinto de Silva, *J. Microsc.* (1971) **12**, 185.
6. Verkleij, A. J., Ververgaert, P. H. J., van Deenen, L. L. M., Elbers, P. F., *Biochim. Biophys. Acta* (1972) **288**, 326.
7. *Ibid.* (1973) **311**, 320.
8. Williams, C. E., Kléman, M., *Proc. 4th Intern. Conf. Liquid Crystals, Stockholm, 1974, J. Phys. Paris* (1975) **36**, C7-375.
9. Kléman, M., *Adv. Liq. Cryst.* (1974) **1**, 210.
10. Lange, Y., Gary-Bobo, C., *J. Gen. Physiol.* (1975) **63**, 640.
11. Ranck, J. L., Mateu, L., Sadler, D. M., Tardieu, A., Gulik-Krzywicki, T., Luzzati, V., *J. Mol. Biol.* (1974) **85**, 249.
12. Steers, M., Kléman, M., Williams, C. E., *J. Phys. Paris Lett.* (1974) **35**, L-21.
13. Friedel, G., Grandjean, F., *Bull. Soc. Fr. Mineral.* (1910) **33**, 409.
14. Bouligand, Y., *J. Phys. Paris* (1972) **33**, 525.
15. Friedel, J., Kléman, M., "Fundamental Aspects of Disclination Theory," *Nat. Bur. Stand. U.S. Spec. Publ.* (1970) **1**, 607.
16. Brochard, F., de Gennes, P. G., Ramana, *Indian J. Phys.*, Suppl. n1 (1975), 1.
17. Bouligand, Y., *J. Phys. Paris* (1973) **34**, 1011.
18. Tardieu, A., private communication.
19. Colliex, C., Kléman, M., Veyssié, *Intern. Cong. Electron Microsc., 8th, Canberra, 1974.*

RECEIVED November 19, 1974.

6

Dynamic Phenomena in Lyotropic Liquid Crystals—A Brief Review

CLARENCE A. MILLER

Department of Chemical Engineering, Carnegie-Mellon University, Pittsburgh, Pa. 15213

Transport phenomena and other dynamic processes in lyotropic liquid crystals have received relatively little research attention. However, they can be quite important in practice because relatively long times are often required to reach equilibrium when a liquid crystal is present. Moreover, understanding of equilibrium behavior seems to have reached a point where additional work on dynamic phenomena would be productive. Accordingly, the available information on such phenomena is reviewed. It consists mainly of measurements of viscosity, diffusivity, electrical conductivity, and chemical reaction rates in liquid crystalline materials. Some possible areas for future research are identified and discussed briefly.

The study of equilibrium situations has dominated research to date on lyotropic liquid crystals. Much has been learned about structure of the various liquid crystalline phases which occur in systems containing water, one or more amphiphilic compounds, and perhaps an oil. Phase diagrams have been determined for numerous binary and ternary systems and for some quaternary systems. Considerable understanding has been developed on how both phase structure and phase equilibrium depend on temperature, molecular structure of the amphiphilic compounds and oil, and electrolyte concentration in the water. Reviews of such information may be found in other chapters of this volume and elsewhere (*1, 2, 3*).

Important questions about equilibrium situations remain to be answered, and a strong continuing research effort in this direction is essential. Nevertheless, enough is now known about equilibrium to justify a strong parallel effort toward understanding nonequilibrium phenomena—an effort much greater than exists at present.

Dynamic processes are prominent among potential applications of lyotropic liquid crystals. It is known, for example, that emulsion stability is greatly enhanced when liquid crystal forms at the surfaces of the drops (*4*). However, emulsion formation is invariably a dynamic process, involving fluid flow and transport of the surfactant from one liquid to the drop surfaces and the other liquid. Since the time required to reach equilibrium in systems containing liquid crystals can sometimes be quite long—weeks or months—the need for knowledge about dynamics of the process is evident.

Another potential application is in detergency. Certain types of dirt are removed by formation of a liquid crystalline phase containing water, detergent, and dirt (*5, 6*). Of primary interest here are the rate at which the liquid crystal forms and whether or not it forms uniformly over the entire surface of the dirt particle (discussed below).

Liquid crystals are widely believed to be closely related to membranes of living cells and have been used as model systems in studies to understand membrane behavior. Among dynamic processes of interest here are transport of various species across membranes and various motions and deformations of membranes.

Existing knowledge of nonequilibrium phenomena in lyotropic liquid crystals is reviewed in this article. First, a few remarks are made about work on hydrodynamic and electrohydrodynamic instabilities in thermotropic liquid crystals and the possibility of applying it to lyotropic systems. Then available results on transport properties—viscosity, electrical conductivity, and diffusivity—in lyotropic systems are summarized. After a brief section on chemical reactions in liquid crystalline phases, the importance of investigating further the kinetics of growth and dissolution of liquid crystals is discussed. Finally, some comments are made about dynamic phenomena in biological systems. In view of the relatively small amount of attention given so far to nonequilibrium phenomena, a continuing effort is made throughout the article to deal with not only what is known but also with what is unknown and should be investigated. The reader is presumed to have a general idea of the equilibrium properties of liquid crystals reviewed in other chapters of this volume—*e.g.*, the basic structure of the various types of liquid crystals found in lyotropic systems.

Applicability of Work on Hydrodynamic Instabilities of Thermotropic Liquid Crystals

In sharp contrast to the situation for lyotropic systems, dynamic behavior of thermotropic liquid crystals has received considerable research attention. It is useful, therefore, to begin by assessing the implications of this work for lyotropic systems.

An active research area for thermotropic materials is study of electrohydrodynamic instabilities (7). Much work has dealt with thin layers (about 10–100 μm) of nematic liquid crystals confined between parallel plates. The plate surfaces are treated or a magnetic field is applied so that the molecules are initially aligned—*e.g.*, with their long axes parallel to the plates. When a potential difference of sufficient magnitude (often only a few volts) is applied between the plates, cellular convection sets in with some resulting disruption of the initial molecular alignment. At somewhat higher voltages alignment is even further disrupted, and scattering of light passing through the material gives it a turbid appearance. This phenomenon is the basis of certain display devices.

Theoretical analysis indicates that occurrence of such convective instabilities depends on anisotropy of electrical conductivity and dielectric properties in the initial aligned nematic material. That is, conductivity parallel to the direction of alignment must differ from conductivity perpendicular to this direction. Calculation of the stability condition requires knowledge not only of these anisotropic electrical properties but also of anisotropic elastic and viscous properties which oppose disruption of the alignment and flow.

Various other instances of hydrodynamic and electrohydrodynamic instabilities in nematic and, to a lesser extent, smectic liquid crystals have been investigated. No attempt is made here to review this work. For the present discussion, it is sufficient to note that (a) most of the work has dealt with oriented layers having anisotropic properties, and (b) some interesting instabilities arise in oriented layers which do not occur for isotropic materials. An example of the latter is cellular convection in a fluid layer confined between horizontal plates maintained at different temperatures. With an isotropic fluid, convection can arise only if the lower plate is hotter than the upper plate. Then, fluid near the lower plate is less dense and tends to rise while fluid near the upper plate is denser and tends to sink. With an oriented layer, however, convection can arise even when the upper plate is hotter if the anisotropy of thermal conduction properties is of a particular type (8).

One potential application of the work on oriented nematic phases of rodlike molecules is to solutions containing cylindrical micelles. Orientation could be achieved by a shear field or perhaps by an electric field. Gotz and Heckman (9) confirmed the existence of anisotropic electrical conductivity for a concentrated surfactant solution in a shear field. They used their results to show that the solution contained cylindrical rather than platelike micelles. Of course, the magnitude of the electrical conductivity in an aqueous micellar solution should be quite different from that in the nematic phase of an organic material. So the conditions for and types of electrohydrodynamic instabilities could be different as well.

Little work seems to have been done on thin oriented layers of lyotropic liquid crystals although there is one recent report of preparation of such a layer of the lecithin–water lamellar phase (*10*). As indicated by Brochard and de Gennes (*11*), theories of the hydrodynamics of thermotropic smectic materials can be adapted to describe oriented layers of lamellar liquid crystal in lyotropic systems.

A few observations can be made about possible hydrodynamic instabilities in lyotropic liquid crystals. First, the available data on lyotropic materials show viscosities in the range of 1–50 poise (*see* below). The nematic materials in which instabilities have been studied have viscosities around 0.1 poise. Hence, instabilities involving cellular convection should be harder to produce in lyotropic materials. Another point is that oriented layers of materials containing alternating lamellae of lipid and water should exhibit tremendous anisotropy in such properties as electrical conductivity and diffusivity. The anisotropy of transport properties in nematic materials is much less with values of a given property in directions parallel and normal to the aligned molecules rarely differing by more than a factor of 2. The extreme anisotropy of transport in lamellar lyotropic phases could greatly influence the forms of instabilities which occur. Finally, because fluid interfaces act to orient amphiphilic molecules, an oriented lamellar phase bounded by at least one fluid phase could be produced. The possibilities of flow along and deformation of an interface between fluid and liquid crystal offer prospects of altering the form of hydrodynamic instabilities.

Transport Properties of Lyotropic Liquid Crystals

Most nonequilibrium systems are characterized by variation of velocity, temperature, composition, or electrical potential with position and the consequent transport of momentum, energy, mass, or electric charge. Naturally, transport of two or more of these may occur simultaneously. Attention is focused here, however, on situations where only one transport process occurs and a transport coefficient can be calculated from its measured rate. For example, thermal conductivity can be calculated if the rate of energy transport and the temperature variation in the system are measured.

As indicated above, much interest exists in dynamic behavior of thin aligned layers of nematic liquid crystals. It is not surprising to find, therefore, that measurement of the anisotropy of transport properties has been the objective of many studies of thermotropic systems. The literature on anisotropic thermal conductivity in nematic liquid crystals has been reviewed recently by Rajan and Picot (*12*). Among the studies of anisotropic diffusion are those of Yun and Fredrickson (*13*), Blinc *et al.* (*14, 15*), and Zupancic *et al.* (*16*). These references are given mainly as examples of the work on thermotropic systems. The techniques used merit study by those interested in measuring anisotropy of transport properties in lyotropic systems.

Most of the work on transport in lyotropic liquid crystals has dealt not with thin oriented layers but with samples made up of many small regions with varying orientations. Transport coefficients obtained from these experiments are thus averaged over orientation. The macroscopic structure of the liquid crystal can significantly influence transport.

The next three sections summarize existing data on transport coefficients. Thermal conductivity is not mentioned because no measurements of this property in lyotropic liquid crystals appear to have been reported.

Viscosity

Viscosity measurement is a well-known technique of studying colloidal systems. Viscosities for solutions containing liquid crystalline material have been reported during studies of aqueous micellar solutions. The problem in interpreting much of this data is that detailed phase diagrams were not reported with the viscosity data. Hence, it cannot be established whether a particular measurement represents properties of a liquid crystalline phase or of a two-phase or three-phase mixture containing some liquid crystal. Perhaps phase diagrams determined in recent years could be used to provide such information about some existing data.

A question which has attracted some attention over the years is the origin of viscoelastic properties of aqueous surfactant solutions containing added electrolyte (*17, 18, 19*). One mechanism which has been suggested is that micelles in these systems are elongated and form a network structure. Another is that the viscoelastic solutions contain some liquid crystalline material, probably as small particles. It would seem that whether any liquid crystal is present could be resolved by determining the phase diagram and making appropriate rheological measurements on a given system of interest. In any case, rheology of mixtures of isotropic solutions and liquid crystals seems not to have been thoroughly and systematically investigated.

Solyom and Ekwall (*20*) have studied rheology of the various pure liquid crystalline phases in the sodium caprylate–decanol–water system at 20°C, for which a detailed phase diagram is available. Their experiments using a cone-and-plate viscometer show that, in general, apparent viscosity decreases with increasing shear rate (pseudo-plastic behavior). Values of apparent viscosity were a few poise for the lamellar phase (platelike micelles alternating with thin water layers), 10–20 poise for the reverse hexagonal phase (parallel cylindrical micelles with polar groups and water in interior of micelles), and 20–50 poise for the normal hexagonal phase (parallel cylindrical micelles with hydrocarbon chains in micelle interiors; continuous aqueous phase). Interesting rheological behavior was observed for the lamellar phase when water content was about equal to or slightly greater than the maximum water of hydration of the polar groups of the soap and alcohol. Under these conditions, apparent viscosity was much higher during initial stages of shearing than at later times. The authors speculated that with small amounts of unbound water, the micelles achieved some freedom of motion and could form three-dimensional structures not possible with lower water content. For larger amounts of unbound water the very high viscosities during initial shearing were not observed. As discussed below, changes in diffusion behavior have been seen in lamellar phases of other systems for water contents corresponding to the maximum water of hydration. These phenomena certainly merit further investigation.

Other work on lamellar phases has confirmed that they are quite viscous and that very high apparent viscosities—or even yielding behavior—occur during initial deformation at some compositions (*21, 22, 23*). Nixon and Chawla (*21*) found in the water–polysorbate 80–ascorbic acid system that the structure leading to initial yielding was irreversibly broken down by recycling the sample several times over a range of shear rates—the behavior of the recycled material being pseudoplastic.

Hyde *et al.* (*24*) found that the peak apparent viscosity of lamellar liquid crystal fell by about two orders of magnitude as alcohol chain length was reduced from 15 to five in water–alcohol–Teepol systems. However, the complete picture of how various types of amphiphilic compounds and their mixtures influence viscosity is not available. In particular, it is not known under what conditions fairly low viscosities of liquid crystals can be achieved although Hallstrom and Friberg (*22*) report viscosities of about 0.2 poise for some compositions in the water–monocaprylin–tricaprylin system. As indicated previously, low viscosities increase the possibilities for occurrence of hydrodynamic instabilities involving cellular convection.

Even when composition is fixed, viscosity and other rheological properties may depend on the size and arrangement of aligned domains within a sample of liquid crystalline material. No studies of this matter seem to have been made, however. Such structural characteristics do influence electrical conduction and diffusion in liquid crystals, as discussed further below.

Another question which has received little attention is whether a shear field can itself influence the conditions when liquid crystalline material is formed. Does some liquid crystal form because of the orienting effect of shear on cylindrical or flat micelles, in composition ranges different from those indicated by the equilibrium phase diagram? The work of Falco *et al.* (*25*) with the water–hexanol–hexadecane–potassium oleate system suggests that the answer is sometimes yes. Friberg and Solyom (*26*) observed that shearing a particular emulsion caused liquid crystalline material to form at the drop surfaces although no liquid crystal was initially present and the overall system composition was in a region of the phase diagram where no liquid crystal should occur. As indicated previously, whether liquid crystal forms is of great importance for emulsion stability. Further studies are needed to clarify how shear affects the formation and break up of liquid crystalline material.

Electrical Conductivity[1]

Interpretation of some of the early data on electrical conductivity is uncertain for the same reason as mentioned above for viscosity—*viz.*, lack of detailed phase diagrams for the systems studied. Much information can be obtained, however, from the investigations which have been better defined.

Figure 1 shows the results obtained by Francois and Skoulios (*27*) on the conductivity of various liquid crystalline phases in the binary systems water–sodium lauryl sulfate and water–potassium laurate at 50°C. As might be expected, the water-continuous normal hexagonal phase has the highest conductivity among the liquid crystals while the lamellar phase with its bimolecular leaflets of surfactant has the lowest conductivity. Francois (*28*) has presented data on the conductivity of the hexagonal phases of other soaps. She has also discussed the mechanism of ion transport in the hexagonal phase and its similarity to ion transport in aqueous solutions of rodlike polyelectrolytes.

Winsor (*17*) describes how electrical conductivity varies during addition of an alcohol to an aqueous micellar solution containing some solubilized oil. Conductivity initially decreases as mixed (and probably larger) micelles containing both surfactant and alcohol are formed. When liquid crystal (presumably having a lamellar structure) starts to appear in equilibrium with the micellar solution, conductivity decreases even faster. As more alcohol is added, the aqueous solution disappears, only liquid crystal is present, and the conductivity reaches a minimum. Addition of still more alcohol results in the appearance of an oil-continuous micellar solution and an increase in conductivity. Eventually all liquid crystal disappears, the increase in conductivity ceases, and conductivity

[1] Note added in proof. Since this article was written, a review of electrical conduction in liquid crystals has appeared: P. A. Winsor, "Liquid Crystals and Plastic Crystals," Vol. 2, p. 122ff, G. W. Gray and P. A. Winsor, Eds., Ellis Horwood, Chichester, 1974.

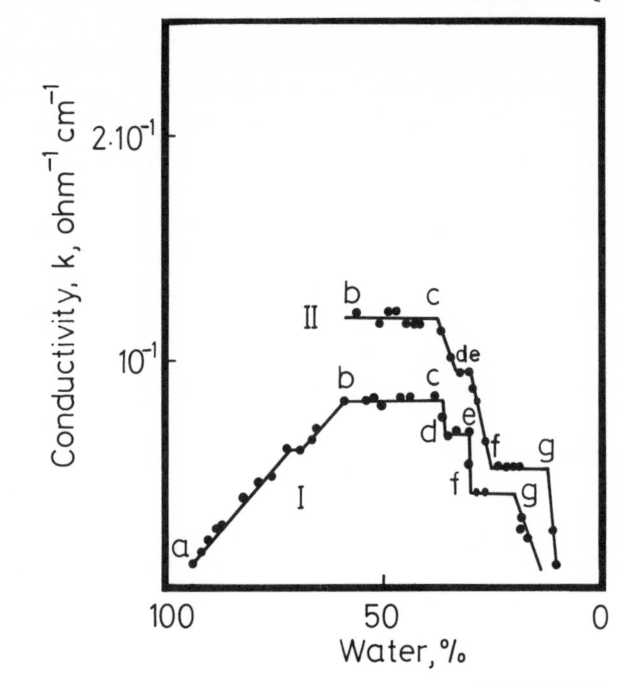

Figure 1. Electrical conductivity of liquid crystalline phases in water–sodium lauryl sulfate (Curve I) and water–potassium laurate (Curve II) systems (27). *Regions ab, bc, de, and fg correspond to micellar solution, hexagonal phase, intermediate phase, and lamellar phase respectively.*

of the oil phase decreases with further alcohol additions. Similar results have been reported by Hyde *et al.* (*24*).

Electrical conductivity in an oriented sample of lamellar phase should be highly anisotropic because ions are far more easily transported along the aqueous layers between the large, flat micelles then across the micelles. For a sample with many different regions or domains of varying orientation, the domains having micelles approximately perpendicular to the applied potential difference provide a significant barrier to current flow. Thus, the relatively low conductivity values found during the work just described seem reasonable. Ekwall *et al.* (*29*) have also measured low conductivities (10^{-5}–10^{-4} ohm^{-1} cm^{-1}) in the lamellar phase of the aerosol OT–water system.

However, the highly anisotropic conductivity within each domain also means that overall sample conductivity is sensitive to features of the macroscopic phase structure. Factors which promote particular arrangements of adjacent domains should influence conductivity. So should structural defects within a domain which would allow appreciable charge transport perpendicular to the micelles. Still another factor is the effect of the applied potential difference itself in orienting domains. Winsor (*17*), Gilchrist *et al.* (*30*), and Francois and Skoulios (*31*) have reported signs that such orientation occurs. Whether the applied field can also influence the conditions when the liquid crystal forms—*i.e.*, whether the phase diagram is field dependent—seems not yet to have been investigated.

Diffusivity

The widespread interest in transport across membranes of living cells has led to studies of diffusion in lyotropic liquid crystals. Biological membranes are generally thought to contain single bimolecular leaflets of phospholipid material, leaflets which are like the large, flat micelles of lamellar liquid crystals. No effort is made here to review the literature on transport either across actual cell membranes or across single bimolecular leaflets (black lipid films) which have often been used recently as model systems for membrane studies. Instead, experiments where lamellar liquid crystals have been used as model systems are discussed.

Several investigators have used radioactive tracer methods to determine diffusion rates. Bangham *et al.* (*32*) and Papahadjopoulos and Watkins (*33*) studied transport rates of radioactive Na^+, K^+, and Cl^- from small particles or vesicles of lamellar liquid crystal to an aqueous solution in which the particles were dispersed. Liquid crystalline phases of several different phospholipids and phospholipid mixtures were used. Because of uncertainties regarding particle geometry and size distribution, diffusion coefficients could not be calculated. Information was obtained, however, showing that the transport rates of K^+ and Cl^- in a given liquid crystal could differ by as much as a factor of 100. Moreover, relative transport rates of K^+ and Cl^- were quite different for different phospholipids. The authors considered that ions had to diffuse across platelike micelles to reach the aqueous phase.

Francois and Varoqui (*34*) measured diffusion rates of Cs^+ in the hexagonal liquid crystalline phases of the water–cesium myristate and water–cesium laurate systems. In each case diffusivity was obtained as a function of temperature for a given liquid crystal composition. Values of $1–2 \times 10^{-5}$ cm²/sec were reported for 60°–80°C. Diffusivity was about an order of magnitude lower in the gel phase of the cesium myristate system.

Axial diffusion of several radioactive ions and molecules in capillary tubes containing the lecithin–water lamellar phase was studied by Lange and Gary Bobo (*35, 36*). As shown in Figure 2, diffusivities increased with increasing water content in the liquid crystal except for a sharp drop

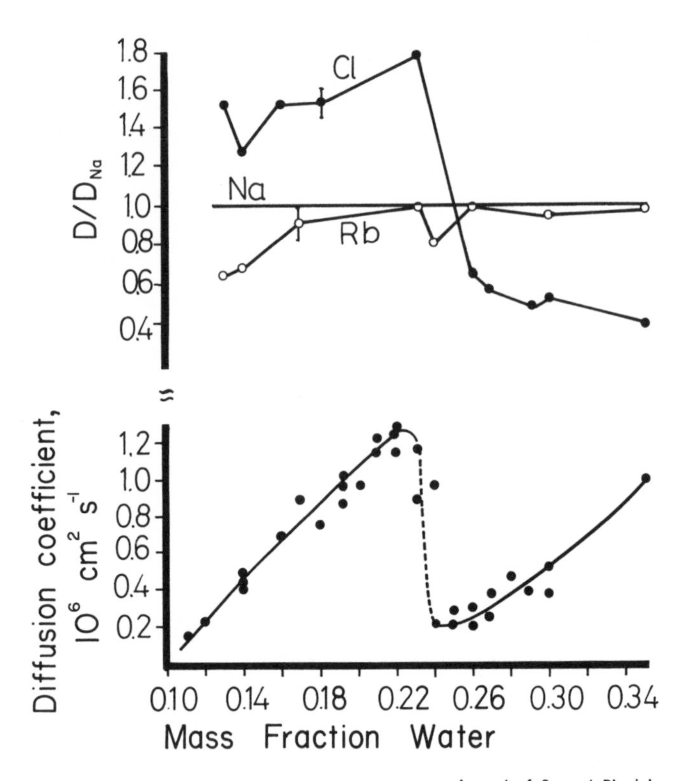

Journal of General Physiology

Figure 2. Diffusion of ions in lecithin–water lamellar phases at 18°C as a function of phase water content, Φ_ω. Upper curve gives diffusion coefficients of Cl^- and Rb^+ relative to that of Na^+. Lower curve gives the diffusion coefficient of Na^+ (36).

at a composition of about 24 wt % water. This composition corresponds to the maximum water which can be tightly bound by polar groups of the lecithin molecules. The relative transport rates of Cl^- and the cations also changed at this composition (*see* Figure 2), and abrupt changes in liquid crystal viscoelastic behavior were noted as well. The authors concluded that some structural change in the phase, possibly a reorientation of the lecithin polar groups, caused the sudden change in properties at this composition. As discussed earlier, Solyom and Ekwall (*20*) saw sudden changes in rheological behavior of the lamellar phase in another system at a composition corresponding to the maximum tightly bound water.

As Figure 2 indicates, ion diffusivities were in the range of 10^{-7} to 10^{-6} cm^2/sec at 18°C. So were diffusivities of benzene and urea, but water diffusivity was somewhat higher, about 0.8–3.0 $\times$ 10^{-6} cm^2/sec. The shape of the diffusivity curves for these molecules was about the same as that given for ions in Figure 2, including the sharp drop at 24% water. These values are effective diffusivities through material with many domains of varying orientation. In view of the relatively small activation energies of diffusion (< 6 kcal/mole) calculated from the variation of diffusivities with temperature, the authors concluded that diffusion in each domain occurred along the thin aqueous layers between the large flat micelles (or within the micelles for benzene).

The diffusion rates of the various ions differed by less than a factor of 3 (*see* Figure 2)—a difference much less than that found in the studies described above for liquid crystalline particles in solution. The reason is presumably that ions must be transported across at least one bimolecular leaflet at the interface between the liquid crystalline particle and the solution. No such interface existed in the experiments of Lange and Gary Bobo.

It is also of interest that the same group of workers found slightly different diffusivities and no sharp drop at 24% water when the liquid crystal was prepared by an earlier procedure (*37, 38*). The authors now feel that some inhomogeneites probably existed in material prepared by the earlier procedure (although results were reproducible). In any case, the different results further emphasize the importance to transport of the macroscopic structure of the liquid crystal.

Devaux and McConnell (*39*) measured a lateral diffusion coefficient of about 2×10^{-8} cm^2/sec at 25°C for phosphatidylcholine (PC) diffusing along bimolecular leaflets in an oriented water–PC lamellar phase. Spin-labeled PC was used in this work.

With spin echo NMR techniques, Charvolin and Rigny (*40*) found a diffusion coefficient of about 2×10^{-6} cm^2/sec for potassium laurate in the cubic liquid crystalline phase it forms with water.

Also using the NMR method, Blinc *et al.* (*14, 41*) report diffusivities of water of 2–8 $\times 10^{-5}$ cm^2/sec at various temperatures in the normal hexagonal phase of the water–sodium palmitate system. At different composition where the lamellar phase exists at these temperatures, their values of diffusivities along the aqueous layers range from 0.7–3 $\times 10^{-5}$ cm^2/sec. These values are about an order of magnitude higher than those given above (*36*), the comparison providing a rough indication of how much diffusion is slowed by the existence of many domains of different orientations.

Chemical Reaction

It might be expected that the ordered arrangement of molecules in liquid crystals could modify the rates of some chemical reactions from those observed in isotropic liquid solutions. Ahmed and Friberg (*42*) studied hydrolysis of *p*-nitrophenyl laurate in the water–hexanol–cetyltrimethylammonium bromide system. They found that the reaction rate was greater by a factor of 2 or 3 in the lamellar liquid crystalline phase than in the isotropic micellar solution. In the hexagonal phase the reaction rate was only slightly above that of the micellar solution. Murthy and Rippie (*43*) had previously discussed a situation where the rate of ester hydrolysis was smaller in the liquid crystalline phase of a nonionic surfactant than in the isotropic solution.

The mechanisms by which liquid crystals influence reaction rates need further study. Presumably, one important factor is the effect of ordering on entropy changes of various steps of the reaction. Bacon (*44*) has considered this factor in connection with the ability of thermotropic liquid crystals to increase the polymerization rate of phenylacetylene.

Kinetics of Liquid Crystal Formation and Dissolution

The results described above show that transport rates in liquid crystals depend not only on the small scale structure of the phase—*e.g.*, whether it is lamellar or hexagonal—but also on aspects of the large scale or macroscopic structure. This conclusion is especially applicable to the lamellar phase with its tremendous degree of anisotropy within a domain of aligned platelike micelles. The arrangement of domains of varying orientation and the existence of structural defects which might facilitate transport perpendicular to the micelles within a domain were mentioned above as being of particular interest. In the extreme case of complete alignment of an entire sample, a material with properties far more anisotropic than seen in thermotropic liquid crystals would exist—a matter of possible interest in its own right.

In view of the importance of macroscopic structure, further studies of liquid crystal formation seem desirable. Certainly, the rates of liquid crystal nucleation and growth are of interest in some applications—in emulsions and foams, for example, where formation of liquid crystal by nonequilibrium processes is an important stabilizing factor—and in detergency, where liquid crystal formation is one means of dirt removal. As noted previously and as indicated by the work of Tiddy and Wheeler (*45*), for example, rates of formation and dissolution of liquid crystals can be very slow, with weeks or months required to achieve equilibrium. Work which would clarify when and why phase transformation is fast or slow would be of value. Another topic of possible interest is whether the presence of an interface which orients amphiphilic molecules can affect the rate of liquid crystal formation at, for example, the surfaces of drops in an emulsion.

However, interest in the process of liquid crystal formation is not

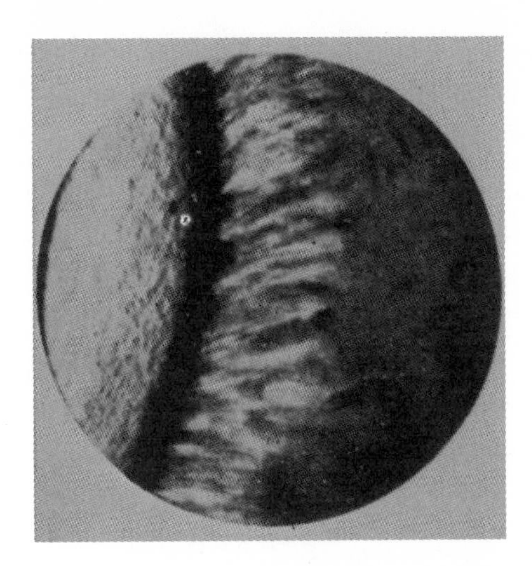

Acta Chemica Scandinavica

Figure 3. Growth of liquid crystalline filaments towards aqueous solution of sodium laurate (right) after it has been contacted with decanol (left) (46)

limited to its rate. The influence of the conditions of phase transformation on the morphological features mentioned above merit further study. The effects of flow, electric fields, and interfaces on orienting domains of aligned micelles could be studied. Even without such effects, Lawrence (*6*) suggests, based on observations of phase texture with a polarizing microscope, that particular arrangements of domains may occur which greatly influence, for example, rheology of the liquid crystal.

Another aspect of growth is morphological stability. Consider the case of detergency where liquid crystal forms at the surface of a dirt particle in contact with a detergent solution. Observations of the process indicate that liquid crystal usually forms not as a layer of uniform thickness, as one might expect, but as a series of filaments which grow toward the aqueous phase (*see* Figure 3). Small particles of liquid crystal break off from the filament tips and are swept away (*5, 46*). The process is evidently closely related to numerous observations over the years of liquid crystal forming as "myelin figures" (*6, 47*). Since a uniform layer of liquid crystal would presumably be less easily removed from the surface of the dirt particle, the formation of liquid crystal as filaments speeds up the detergency process.

The filaments which grow from a dirt particle are reminiscent of dendrites which often form near a surface at which liquid solidifies. The beautiful patterns of snowflakes seen under magnification are dendritic structures, for example. It is now generally believed that when a solidification front does not remain smooth but develops a dendritic structure, the reason is an instability of the surface arising from heat and mass transport effects (*48*). In a similar way, diffusion could be one factor leading to filaments rather than smooth surfaces during formation of liquid crystalline phases (*49*).

While the emphasis of this section has been on kinetics of liquid crystal formation, the rate of liquid crystal dissolution may also be of interest—*e.g.*, in connection with breaking foams and emulsions by adding materials which destroy the liquid crystalline structure.

Biological Systems

An exciting potential application of knowledge about lyotropic liquid crystals is in biological systems. Another chapter in this volume discusses biological systems as its chief subject. Here only a few remarks are made about dynamic phenomena.

The close relationship between cell membranes and liquid crystals has already been cited as a stimulus for research on transport of ions and various uncharged molcules in liquid crystals. However, dynamic behavior of membranes is not limited to transport processes. Ambrose (*50*) has provided a fascinating description of cell surfaces in constant motion with pseudopodia continually forming, growing, and disappearing. He also noted that such behavior is quite different for normal and cancer cells, at least for the cell types he studied.

The overall cell surface region is generally considered to have both a bimolecular leaflet containing primarily phospholipids and an underlying "plasma gel" layer containing many long microfibrils. Alignment of the microfibrils is observed when the cell surface extends to form pseudopodia, according to Ambrose. Under these conditions, the plasma gel closely resembles an oriented layer of liquid crystal. The microfibril alignment presumably influences the stress distribution within the cell surface region and may thus aid in causing deformation of the cell surface. Some preliminary thoughts about mechanical properties of cell surfaces have been advanced by Fergason and Brown (*51*) and Ambrose (*50*), but further work is needed.

Deposition of liquid crystalline material containing cholesterol esters on the inner surfaces of human arteries is another nonequilibrium process in which much interest exists. No doubt there are other examples as well where dynamic phenomena involving liquid crystals are important in biological systems.

Conclusions

Although dynamic behavior of lyotropic liquid crystals is important in various applications, relatively little work has been done on the subject.

The former factor explains why an article on dynamic phenomena is included in this volume; the latter explains why the article is as much a survey of topics for further research as it is a review of past work. Prospects of exciting new research developments in this field seem bright.

Literature Cited

1. Skoulios, A., *Adv. Colloid Interface Sci.* (1967) **1,** 79.
2. Winsor, P., *Chem. Rev.* (1968) **68,** 1.
3. Ekwall, P., Danielsson, I., Stenius, P., "Surface Chemistry and Colloids," p. 97 ff, M. Kerker, Ed., Butterworths, London, 1972.
4. Friberg, S., Mandell, L., Larsson, M., *J. Colloid Interface Sci.* (1969) **29,** 155.
5. Stevenson, D., *J. Textile Inst.* (1953) **44,** T12.
6. Lawrence, A. S. C., "Surface Activity and Detergency," K. Durham, Ed., p. 158 ff, Macmillan, London, 1961.
7. Helfrich, W., *Mol. Cryst. Liq. Cryst.* (1973) **21,** 187.
8. Dubois-Violette, E., *C.R. Acad. Sci. Paris* (1971) **B273,** 923.
9. Gotz, K., Heckman, K., *Disc. Faraday Soc.* (1958) **25,** 71.
10. Powers, L., Clark, N. A., *Abstract, Intern. Liq. Cryst. Conf., 5th, Stockholm, 1974.*
11. Brochard, F., de Gennes, P., *Abstract, Intern. Liq. Cryst. Conf., 5th, Stockholm, 1974.*
12. Rajan, V., Picot, J., *Mol. Cryst. Liq. Cryst.* (1972) **19,** 55.
13. Yun, C. K., Fredrickson, A. G., *Mol. Cryst. Liq. Cryst.* (1970) **12,** 73.
14. Blinc, R., Dimic, V., Pirs, J., Vilfan, M., Zupancic, I., *Mol. Cryst. Liq. Cryst.* (1971) **14,** 97.
15. Blinc, R., Zupancic, I., Pirs, J., Luzar, M., Jamsek, M., Doane, J., *Abstract, Intern. Liq. Cryst. Conf., 5th, Stockholm, 1974.*
16. Zupancic, I., Pirs, J., Luzar, M., Blinc, R., Doane, J., *Abstract, Intern. Liq. Cryst. Conf., 5th, Stockholm, 1974.*
17. Winsor, P., "Solvent Properties of Amphiphilic Compounds," Butterworths, London, 1954.
18. Booij, H. L., Bungenberg de Jong, H. G., "Protoplasmatologia, Handbuch der Protoplasmaforschung," Vol. 1, No. 2, Springer Verlag, Vienna, 1956.
19. Pilpel, N., *Trans. Faraday Soc.* (1966) **62,** 1015.
20. Solyom, P., Ekwall, P., *Rheol. Acta* (1969) **8,** 316.
21. Nixon, J. R., Chawla, B., *J. Pharm. Sci.* (1967) **19,** 489.
22. Hallstrom, B., Friberg, S., *Acta Pharm. Suecica* (1970) **7,** 691.
23. Tamamushi, B., *Abstract, Intern. Liq. Cryst. Conf., 5th, Stockholm, 1974.*
24. Hyde, A. J., Langbridge, D. M., Lawrence, A. S. C., *Disc. Faraday Soc.* (1954) **18,** 239.
25. Falco, J. W., Walker, R. D., Shah, D. O., *AIChE J.* (1974) **20,** 510.
26. Friberg, S., Solyom, P., *Kolloid-Z.* (1970) **236,** 173.
27. Francois, J., Skoulios, A., *Kolloid-Z.* (1967) **219,** 144.
28. Francois, J., *Kolloid-Z.* (1971) **246,** 606.
29. Ekwall, P., Mandell, L., Fontell, K., *J. Colloid Interface Sci.* (1970) **33,** 215.
30. Gilchrist, C. A., Rogers, J., Steel, G., Vaal, E. G., Winsor, P., *J. Colloid Interface Sci.* (1967) **25,** 409.
31. Francois, J., Skoulios, A., *C.R. Acad. Sci. Paris* (1969) **C269,** 61.
32. Bangham, A. D., Standish, M. M., Watkins, J. C., *J. Mol. Biol.* (1965) **13,** 238.
33. Papahadjopoulos, D., Watkins, J. C., *Biochim. Biophys. Acta* (1967) **135,** 639.

34. Francois, J., Varoqui, R., *C.R. Acad. Sci. Paris* (1968) **C267,** 517.
35. Lange, Y., Bobo, C. Gary, *Nature New Biol.* (1973) **246,** 150.
36. Lange, Y., Bobo, C. Gary, *J. Gen. Physiol.* (1974) **63,** 690.
37. Rigaud, J., Bobo, C. Gary, Lange, Y., *Biochim. Biophys. Acta* (1972) **266,** 72.
38. Lange, Y., Bobo, C. Gary, Soloman, A., *Biochim. Biophys. Acta* (1974) **339,** 347.
39. Devaux, P., McConnell, H., *J. Am. Chem. Soc.* (1972) **94,** 4475.
40. Charvolin, J., Rigny, P., *J. Chem. Phys.* (1973) **58,** 3999.
41. Blinc, R., Easwaran, K., Pirs, J., Volfan, M., Zupancic, I., *Phys. Rev. Lett.* (1970) **25,** 1327.
42. Ahmed, S., Friberg, S., *J. Am. Chem. Soc.* (1972) **94,** 5196.
43. Murthy, K. S., Rippie, E. G., *J. Pharm. Sci.* (1970) **59,** 459.
44. Bacon, W. E., *Abstract, Intern. Liq. Cryst. Conf., 5th, Stockholm, 1974.*
45. Tiddy, G., Wheeler, P., *Abstract, Intern. Liq. Cryst. Conf., 5th, Stockholm, 1974.*
46. Ekwall, P., Salonen, M., Krokfors, I., Danielsson, I., *Acta Chem. Scand.* (1956) **10,** 1146.
47. Dervichian, D. G., *Progr. Biophysics Mol. Biol.* (1964) **14,** 263.
48. Sekerka, R., *J. Cryst. Growth* (1968) **3/4,** 71.
49. Miller, C. A., *Abstract, Intern. Liq. Cryst. Conf., 5th, Stockholm, 1974.*
50. Ambrose, E. J., *Symp. Faraday Soc.* (1971) **5,** 175.
51. Fergason, J. L., Brown, G. H., *J. Am. Oil Chem. Soc.* (1968) **45,** 120.

RECEIVED November 19, 1974.

7

Some Aspects of Liquid Crystal Microdynamics

JEAN CHARVOLIN

Laboratoire de Physique des Solides[1],
Bât. 510, Université Paris-Sud, 91405 Orsay, France

Interest in thermotropic liquid crystals has focussed mainly on macroscopic properties; studies relating these properties to the microscopic molecular order are new. Lyotropic liquid crystals, e.g. *lipid–water systems, however, are better known from a microscopic point of view. We detail the descriptions of chain flexibility that were obtained from recent DMR experiments on deuterated soap molecules. Models were developed, and most chain deformations appear to result from intramolecular isomeric rotations that are compatible with intermolecular steric hindrance . The characteristic times of chain motions can be estimated from earlier proton resonance experiments. There is a possibility of collective motions in the bilayer. The biological relevance of these findings is considered briefly. Recent similar DMR studies of thermotropic liquid crystals also suggest some molecular flexibility.*

In thermotropic liquid crystal studies, great effort was expended in investigating macroscopic properties such as order, elasticity, and viscosity. These properties depend, of course, on microscopic parameters that are characteristic of the interactions between individual molecules. Since most mesogen molecules are very complex structurally, these interactions are expected to depend greatly on the molecular conformation. Nevertheless, thermotropic liquid crystals have been analyzed mainly as assemblies of rigid cylindrical rods. Studies that attempt to relate macroscopic and microscopic orders are just beginning. The preliminary findings from such studies were discussed at the Fifth International Liquid Crystal Conference; they indicate that the traditional rod picture is not adequate for a real understanding of most liquid crystalline phases. This point of view

[1] Laboratoire associé au CNRS.

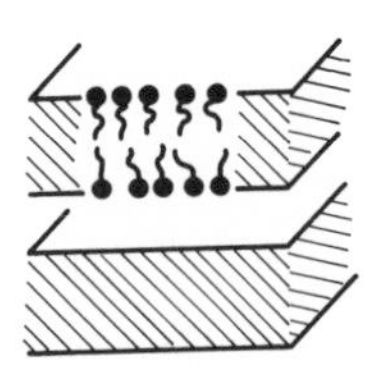

Figure 1. Schematic representation of two lyotropic mesophases. The lamellar phase (left) is a periodical stacking along one dimension of soap and water lamellae. In the hexagonal phase (right), the soap cylinders are organized in a two-dimensional array.

is reinforced by recent theoretical considerations that stress the importance of end chains in the thermodynamics of smectic (*1*) and nematic (*2*) mesophases.

It is interesting that work on the internal motions of the molecules that produce lyotropic mesophases is more advanced. This is mainly because of the importance of the microscopic properties of these systems in solubilization and interfacial problems, problems which are encountered in industry as well as in cell membrane biology. The structural and functional roles of lipid molecules in biomembranes are much discussed; investigations of the physicochemical properties of lipid media thus might provide orientations for biological studies. Moreover, the findings on the flexibility of the paraffinic chains in lyotropic mesophases might also be relevant to similar problems in thermotropic mesophases.

Paraffinic Chain Flexibility in Bilayers

Lipid–water systems are the simplest of lyotropic liquid crystals. Because of the strong amphipatic character of lipids, these systems constitute distributions of two media: the aqueous and the paraffinic. The much diversified geometry of the distributions was demonstrated by x-ray (*3*). To summarize briefly, the molecular aggregates can be lamellae, ribbons, rods, or spheres organized in long range lattices with periodicities in one, two, or three dimensions. The simplest way to understand this rich polymorphism is to assume that the paraffinic chains are flexible in order to fill uniformly the irregularly shaped volumes that are available to them. This idea is compatible with the short range disorder detected at the molecular level by wide-angle x-ray scattering (*3*). The schematic representations of two lyotropic mesophases (Figure 1) depict the coexistence of a long range order with a short range disorder. We shall focus our interest on some studies of paraffinic chain behavior in bilayer lamellar structures.

Magnetic Resonance Studies. Several experimental techniques that are sensitive to local fluctuations have been used. Of these, nuclear magnetic resonance (NMR) yields, in this particular field, the most reliable quantitative data. In the past, most NMR studies, lyotropic as well as thermotropic, were of the protons that are naturally present in the molecules. However, in such anisotropic systems, the residual dipolar coupling between the protons are very large, all nuclear spins are coupled, and it is impossible to resolve specific molecular groups (*4*). A decisive step was the NMR study of deuterated molecules (DMR). Since the dipolar couplings between deuterons are negligible, the system is one of uncoupled spins. Different molecular groups can be distinguished by differences in the motional averaging of the quadrupolar couplings of their C–D bonds.

In the presence of a quadrupolar coupling, the NMR line of a deuteron is split into a symmetric doublet. With an axial electric field gradient (e.f.g.), the doublet spacing in frequency units is given by:

$$\Delta\nu = \frac{3}{4}\frac{e^2qQ}{h}(3\cos^2\theta - 1) \tag{1}$$

where e^2qQ/h is the static quadrupolar coupling constant and θ is the angle between the e.f.g. axis and the magnetic field. Motions of this axis modulate θ, and they should be time-averaged if the motion frequencies are larger than the static interaction. If the motion is anisotropic and admits an axis of symmetry, Equation 1 becomes:

$$\Delta\nu = \frac{3}{4}\frac{e^2qQ}{h} < \frac{3\cos^2\theta' - 1}{2} > (3\cos^2\phi - 1) \tag{2}$$

where θ' and ϕ are the angles made by the symmetry axis with the e.f.g. axis and the magnetic field, respectively. An effective quadrupolar coupling constant is measured which is:

$$\frac{e^2qQ}{h} < \frac{3\cos^2\theta' - 1}{2} > \quad \text{where} < \frac{3\cos^2\theta' - 1}{2} >$$

can be defined as the order parameter, S, of the axis of the e.f.g. relative to the axis of symmetry.

A spectrum of perdeuterated potassium laurate in oriented soap–water multilayers (*5*) is presented in Figure 2. The measured quadrupolar coupling constants are much smaller than those of a static C–D bond, which are about 167 kHz (*6*); the residual quadrupolar splittings

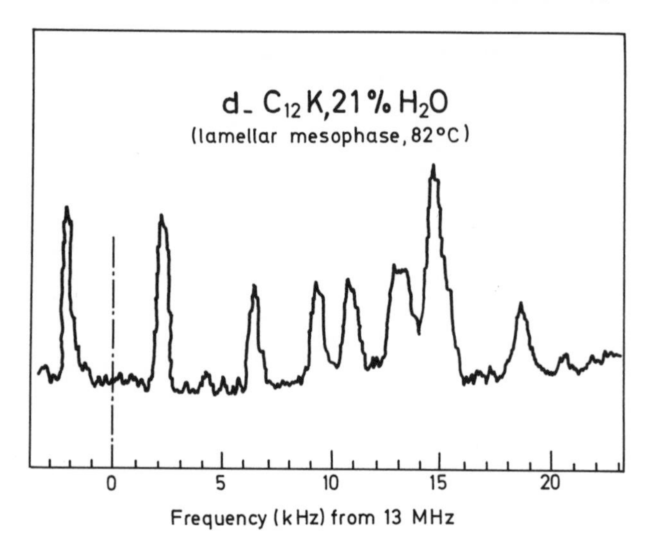

Figure 2. A DMR spectrum obtained at 13 MHz with an oriented sample. The normal to the lamella is at $\pi/2$ from the external magnetic field. Only half the spectrum, which is symmetric about the zero central frequency, is pictured. Each line can be attributed to a methylene, or methyl, group; the greater the shift, the closer the group to the polar head.

reveal that anisotropic intramolecular motions take place and average much of the static coupling. The measured splittings are thus proportional to the order parameters S of the C–D bonds with respect to the optical axis of the liquid crystal. Since the splittings are distinct, the amplitudes of the motions must vary all along the chain. Variation in the splitting with location of the methylene on the chain is plotted in Figure 3. Seelig and Niederberger (7) deuterated specifically each methylene group of sodium decanoate and observed very similar variation in a ternary lamellar system (sodium decanoate–decanol–water). Such a curve can be considered an accurate representation of chain flexibility in the paraffinic medium.

Chain Deformations. A striking feature of the splittings in Figure 3 is that—excluding the first, second, and last three segments of the chain—all the other methylene groups have nearly the same order parameter. For an isolated chain, this would be very odd because each segment is expected to have greater orientational freedom than the preceding one. In other words, the flexibility of a free chain could not yield such a plateau. However, for a chain in a bilayer, the steric repulsions between neighboring chains should also be considered; they result from the constraints imposed by the packing of the lipid polar heads at the interface.

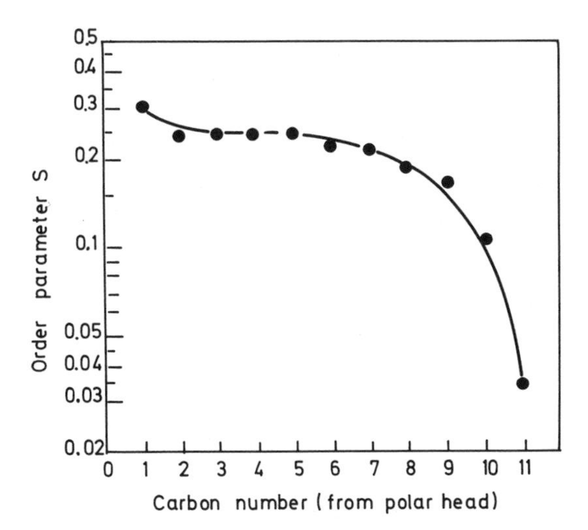

Figure 3. Variation in the order parameter S *of a C–D bond as the distance from the polar head increases.* S *is defined relative to the symmetry axis of the lamella.*

Recently, de Gennes considered the situation in which the mean spacing between chains, when measured normally to the chain elongation axis, remains uniform in the regions of the bilayers without chain ends. He demonstrated that the steric interactions constrain the chains in such a way that orientational disorder cannot increase along the chain as it would for an isolated chain (*8*). The decrease of the order parameters for the last few segments is then explained by the presence of many chain ends in the central region of the bilayer which makes the steric hindrance much less stringent. If this interpretation is correct, then the fact that this drop involves the last three segments of the chain implies that the chain ends are not confined to a plane in the exact center of the bilayer, but rather that they are distributed over a region of sizeable thickness. This, in turn, implies that the chain deformations are not restricted to small angle torsions or flexions of the bonds, but they also involve isomeric rotations around the C–C bonds which can move the chain end appreciably. A free chain with n links has about 3^n rotational isomers (*9*) many of which are certainly still accessible in the mesophase. To describe in detail the thermal motions between all these conformations would be an overwhelming task. However, a few very specific models can be developed if one considers only the simplest isomers.

A chain subgroup has three rotational isomers because of rotations around the C–C bond (*9*): one trans (t) isomer which is depicted in the structural sketch and two gauche (g^+ and g^-) isomers that have higher

Chain subgroup

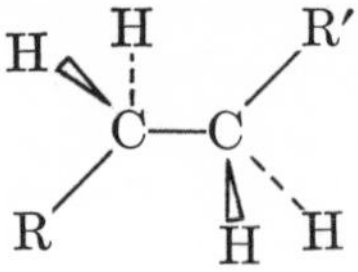

energies than the trans isomer ($\Delta E \sim 500$ cal/mole). For a chain, departures from the low-energy all-trans state can be named according to the succession of the chain C–C bond conformations. An isolated gauche conformation has a $tg^{+}t$ or a $tg^{-}t$ sequence, and a kink (*10*) can be written as $tg^{+}tg^{-}t$ or $tg^{-}tg^{+}t$. Trans sequences on both sides of a kink remain parallel to each other; they are shifted only laterally.

In reference to modeling, the occurrence of isolated gauche conformations with equal probability all along the chain yields an exponential decrease in the order of the methylene along the chain. As was stated above, this is ruled out by the DMR measurements. We conclude that this model could represent the behavior of an isolated chain fixed at one end but not that of a chain in a bilayer since it does not take into account the steric repulsions between neighboring chains. Seellig and Niederberger (*7*) did consider this factor; they combined two gauche isomerizations of different sign into a kink (*10*) that migrates along the chain. This mobile defect decreases equally the order of all methylene groups, leaving the two parts of the chain on either side of it parallel to the bilayer normal. Quantitative agreement with Figure 3 is good if one assumes a probability $p_t \simeq 0.8$ for a link to be in a trans state. In this one-chain model, neighboring molecules are introduced implicitly because one is restricted to defects which do not change the overall direction of the chain. Other defects of this type, with different ranges along the chain (*11*), could be considered as well. Such models, which introduce intermolecular cooperativity arbitrarily by eliminating presupposed unfavorable conformations, skip the interesting aspect of possible collective deformations of lipids in bilayers.

Several approaches to the many-chains problem have been proposed: detailing two-chain interactions (*12*), solving exact two-dimensional lattice models (*13*), and calculating a self-consistent molecular field approximation (*14*). In this last approach, Marčelja extended to lyotropic liquid crystals some of the ideas he developed for thermotropic systems (*2*). More specifically, the packing of the lipid polar heads at the interface is introduced through a lateral pressure term which is estimated from surface pressure measurements in monolayers. Statistical averages are then evaluated by summation over all conformations of a chain in the field that result from neighboring molecules. The calculated behavior of the order of the chain links agrees with Figure 3 and Ref. 7. A palmitate chain (C_{16}) would then have four links in gauche conformations which agrees with earlier thermodynamic calculation (*15*). This theory also gives good agreement with the measured thermodynamic properties of the transition where all the chains are in a rigid, all-trans conformation.

It is interesting to note that spin label experiments reveal a monotonic decrease in the order from head to tail (*7*) which contrasts with the previously discussed plateau that is obtained by deuteron experiments. Although it is not possible at the moment to eliminate the idea of perturbing effects from the nitroxide label, it is nevertheless interesting to attempt to compare EPR and DMR data only in terms of lipid behavior. Recently, Gaffney and McConnell (*16*) suggested that one could reconcile the two types of experiments by taking into account their different time scales. A motion is said to be fast if its frequency is larger than the frequency of the interaction it modulates; this is the so-called motional-narrowing condition. Typical values for the interactions are 10^8 Hz in EPR and 10^5 Hz in DMR. A motion which has a frequency of 10^5–10^8 would appear as a slow motion in EPR (a distributed spectrum is observed) and as a fast motion in DMR (a narrowed spectrum is observed). In the Gaffney and McConnell model (*16*), local fast motions, with frequencies greater than $\sim 10^8$ Hz, are responsible for the monotonic decrease in the spin label order parameters, and each methylene group occupies a larger effective volume than the preceding one. The region near the interface is then a weak density region. A plausible solution to this packing problem is a collective bending of the chains: their elongation axis, which is normal to the bilayer in the tail region, is tilted in the vicinity of the interface (*17*). This model accounts for the EPR spectra if one assumes a static bending angle, or, to express it differently, if one assumes that the bilayer has local biaxiality at times $> 10^{-8}$ sec; the failure of DMR experiments to detect any biaxiality in potassium laurate bilayers (*5*) can be explained by the fact that the lifetime of this biaxiality is 10^{-5}–10^{-8} sec. This tilt angle, which affects the first methylene groups, would then introduce a reducing factor into the expressions of their DMR splittings, thereby transforming into a plateau the monotonic decrease in their order that is caused by fast motions. In this model, contrary to that of de Gennes, the mean lateral spacing is not constant through the bilayers. As has been stated, this assumption is supported mostly by the need for a bending angle as a fitting parameter for the EPR spectra (*17*). Nevertheless, it remains a description of the behavior of labelled chains which, in the absence of direct conclusive experiments, can still be considered somewhat different from that of nonlabelled molecules demonstrated by recent experiments in the thermotropic field (*18*). (Order parameters and tilt angles determined by spin label experiments in Sm C phases depend on the location of the label on the alkoxy chain of the molecule.)

Dynamics of the Deformations. A complete description must also provide the characteristic times of the motions and their activation energies. This can be obtained from NMR experiments through the modulation of the spin Hamiltonian by the motions. If $J(\omega)$ is the spectral density associated with a fluctuating spin Hamiltonian, the relaxation rate of the Zeeman energy at frequency ω can be approximated by $T_1^{-1} \propto J(\omega)$. For example, the spectral density of a fluctuating magnetic field with an exponential correlation function would be $<h^2> 2\tau_c(1 + \omega^2\tau^2_c)^{-1}$, where $<h^2>$ is the mean quadratic strength of the interaction, and τ_c is the correlation time of its variations. Measurement of the relaxation rate as a function of frequency, in the laboratory frame (T_1^{-1}) and in the rotating frame $(T_{1\rho}^{-1})$, provides information about the correlation functions of the motions in a dynamic range from a few Hertz to 10^8 Hz.

In this field, the resolution of DMR is promising. However, experiments on deuterated molecules have just begun, and the nuclear relaxation was not yet analyzed. We can just present here some preliminary ideas that were obtained from proton relaxation experiments (*19*). Because of the nature of dipolar interaction, we are dealing with a multispin system; this entails some complex problems of nuclear spin dynamics which are beyond the scope of this discussion. The quantitative analysis of proton relaxation data is thus far from straight-forward (*20*). We shall limit ourselves to a qualitative interpretation of the frequency dependence of the relaxation rate that is summarized schematically in Figure 4. Important relaxation effects appear in both high and low frequency regions.

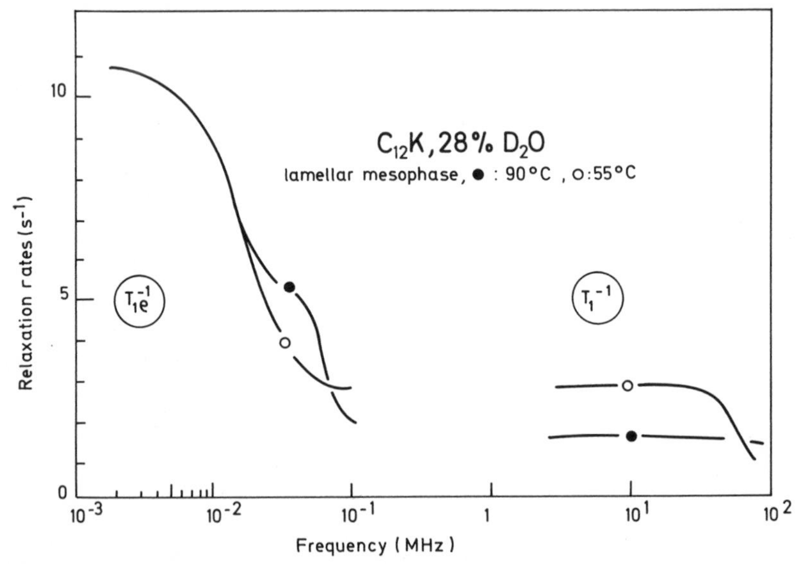

Figure 4. Schematic representation of proton relaxation rates as function of frequencies. No measurements were made in the 0.1 MHz range which is between the domain of the $T_{1\rho}$ *and* T_1 *techniques.*

The high frequency relaxation is attributed in part to the modulation of intermolecular dipolar interactions by the translational diffusion. The cutoff frequency (60 MHz at 55°C) corresponds to the local diffusive jump frequency that is estimated from measurements of the diffusion coefficient ($D \sim 10^{-6}$ cm^2/sec at 55°) (*19, 21*). This cutoff frequency also varies in temperature with the same activation energy ($E_{act} \sim 0.25$ eV) as the diffusion frequency.

The low frequency relaxation can then be attributed to the modulation of intra- and intermolecular interactions by the deformations of the molecules. In this region, the $T_{1\rho}^{-1}$ curves are not simple Lorentzians as they should be if only one motion were responsible for the relaxation. They can be analyzed as a superposition of Lorentzian curves with different characteristic frequencies whose lower limit is about 10^5 Hz. This suggests that the chain deformations result in the appearance of a distribution of modes, each with its own lifetime. The slowest mode has a temperature-independent characteristic time of 3.5×10^{-6} sec; it could be a long wavelength deformation of the molecule. We cannot derive accurate determinations for the fastest local modes. A theoretical consideration by Ågren (*22*), that develops a Rouse analysis familiar in the field of polymer physics (*23*), estimates their characteristic times at about 10^{-8} or 10^{-9} sec; these could interfere with translational diffusion in the high frequency relaxation range. One must remark that, since each motion modulates several interactions of different origin, it is effectively hopeless to try to interpret the values of the modulated interactions that are derived from relaxation data. Consequently, rely only on the frequency and temperature dependences of the relaxation rates and never on their absolute values.

Discussion. We can now propose a coarse description of the paraffinic medium in a lamellar lyotropic mesophase (potassium laurate–water). Fast translational diffusion, with $D \sim 10^{-6}$ at 90°C, occurs while the chain conformation changes. The characteristic times of the chain deformations are distributed up to 3.10^{-6} sec at 90°C. Presence of the soap–water interface and of neighboring molecules limits the number of conformations accessible to the chains. These findings confirm the concept of the paraffinic medium as an anisotropic liquid. One must also compare the frequencies of the slowest deformation mode (10^6 Hz) and of the local diffusive jump (10^9 Hz). When one molecule wants to slip by the side of another, the way has to be free. If the swinging motions of the molecules, or their slowest deformation modes, were uncorrelated, the molecules would have to wait about 10^{-6} sec between two diffusive jumps. The rapid diffusion could then be understood if the slow motions were collective motions in the lamellae. In this respect, the slow motions could depend on the macroscopic structure (lamellar or cylindrical, for example) whereas the fast motions appear to be qualitatively independent of it and governed essentially by the density of the medium (*19, 24*). Unfortunately, no other investigation besides ours has been done in the low frequency region so far.

As is reported in Tiddy's recent, well documented review (*25*), more and more relaxation experiments are concerned with lipids of biological significance in vesicular structures which approximate biomembranes more than our lamellar lyotropic liquid crystal can. Because of the complexity of the molecules and the somewhat uncertain state of determining structural parameters of vesicles, the interpretation of these data is sometimes less straightforward than in the case of simple lyotropic liquid crystals. Nevertheless, such physical studies reveal that, in addition to their rich polymorphism, lipid–water systems exhibit complex dynamic properties that are associated with molecular deformations and fluidity. The lipid medium thus appears capable of fast, local, structural transformations in response to external physicochemical stimulation. Transport properties in and across the bilayer, for example, are related to the chain transition from an ordered to a disordered state that is induced by temperature variation. Whatever interest such phenomena might present in the interpretation of biomembrane properties (*26*), nothing can yet be said about their importance in biological functions. The bilayer is an extreme oversimplification, a fragmentary representation of actual membranes, and much is to be expected from complex lipids and proteins as well as from lipid-protein interactions.

Alkyl Chain Behavior in Smectic Phase

The deuteron spectrum obtained from a thermotropic smectic A phase is presented in Figure 5. The butoxybenzylidene-*p*-octylaniline molecule (BBOA) was synthesized after deuteration of the octyl aniline group. From the center (zero frequency) of the spectrum, there appear successively: a methyl line, a partially deuterated phenyl group doublet, intermediate methylene lines, and the line of the methylene close to the benzyl group. These findings and their relation to previous calculation (*2*) are presented and discussed in a preliminary report (*27*). At this time, liquid crystal studies appear to be at a cross-point. Study of the microdynamics of thermotropics can benefit from the more advanced state of this work for lyotropics, whereas the experience gained in the optical macroscopic studies of thermotropics is useful in the lyotropic field for investigating collective motions such as those discussed above.

Conclusion

Some recent DMR and EPR experiments with lyotropic liquid crystals have been described briefly and interpreted in terms of the flexibility of

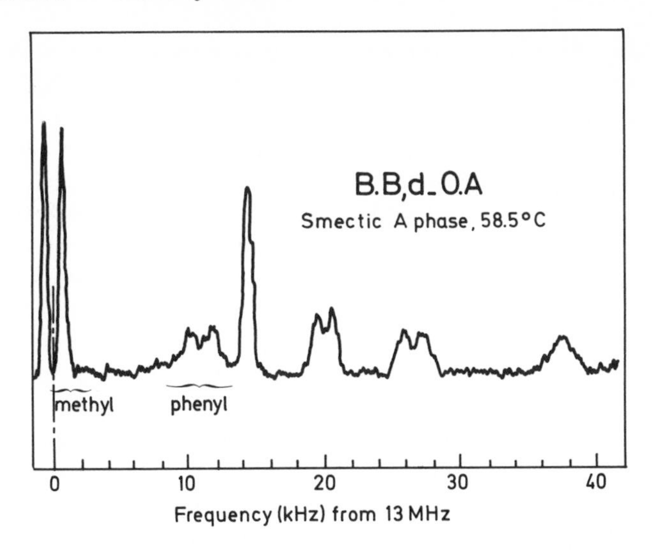

Figure 5. A DMR spectrum obtained at 13 MHz with the thermotropic sample BBOA in the smectic A phase. Only half the spectrum, which is symmetric about the zero frequency, is pictured.

paraffinic chains. Some findings about lateral diffusion of the molecules in the bilayer were cited. The characteristic times of these motions were discussed on the basis of previous proton relaxation experiments. This set of data permits a preliminary description of the lipid medium. Comparisons with liquid paraffins could be attempted: the densities do not differ drastically (The chain disorders in lyotropic liquid crystals and in liquid paraffins are characterized by analogous x-ray diffuse bands at 4.5 A^{-1}.) and chain deformations resulting from reorientations around the C–C bonds are also present (*28*). However, an original and fundamental difference is introduced by the packing of the polar heads at the interface: lipid diffusion along the direction normal to the interface is of very low probability, and chain ends are localized either at the interface or in the central region of the bilayer. Thus, there is less free volume in the bulk of each half bilayer than there is in liquid paraffins where the chain ends are distributed uniformly. This greater lateral cohesion could be an important factor in the development of long-wavelength collective modes in lipid–water systems, the existence of which is suggested by our dynamic data. (A theoretical approach to collective modes in lipid–water systems was presented at the Fifth International Liquid Crystal Conference. In addition to the usual smectic modes, which depend on the symmetry of the system, an extra hydrodynamic mode appears that is related to the presence of two components, *i.e.* to fluctuations in the relative concentrations of water and lipids (*29*).)

However, if reference is made to thermotropic studies, collective motions in smectic phases have wavelengths that extend up to the optical wavelength. In this respect, they were investigated better by light-scattering techniques than by a technique as local as NMR. Finally, nothing has been said about molecular rotation around the long axis, a domain where local techniques can still be useful. With a flexible disordered chain, it is difficult to distinguish clearly between the relative reorientations of its segments and its overall rotation; on the other hand, if bilayers with tilted chains in trans conformations are considered (*30*), it would be of great interest to determine if their macroscopic biaxiality proceeds from any anisotropy in the overall rotation of the molecule. A similar problem arises in the thermotropic field with understanding the biaxial smectic C and H phases. This point was much discussed at the Fifth International Liquid Crystal Conference. It was demonstrated recently for TBBA in Sm H phase that no anisotropy in the overall rotation of the molecule around its long axis is apparent at times longer than about 10^{-10} sec (*31*).

Acknowledgment

Most of our proton NMR studies on lipid–water systems were made in P. Rigny's laboratory (Division de Chimie, C.E.N. Saclay). Many of the ideas that are expressed here were discussed under his critical and stimulating guidance.

Literature Cited

1. McMillan, W. L., *Phys. Rev. A* (1971) **4**, 1238.
2. Marčelja, *J. Chem. Phys.* (1974) **60**, 3599.
3. Luzzati, V., in "Biological Membranes," D. Chapman Ed., Academic, New York, 1968.
4. Charvolin, J., Rigny, P., *Nat. New. Biol.* (1972) **237**, 127.
5. Charvolin, J., Manneville, P., Deloche, B., *Chem. Phys. Lett.* (1973) **23** 345.
6. Burnett, L. J., Muller, B. M., *J. Chem. Phys.* (1971) **55**, 5829.
7. Seelig, J., Niederberger, W., *Biochemistry* (1974) **13**, 1585.
8. de Gennes, P. G., *Phys. Lett. A* (1974) **47**, 123.
9. Flory, P. J., "Statistical Mechanics of Chain Molecules," Interscience, New York, 1969.
10. Pechhold, W., *Kolloid Z. Z. Polym.* (1968) **228**, 1.
11. Dubois-Violette, E., Geny, F., Monnerie, L., Parodi, O., *J. Chem. Phys.* (1969) **66**, 1865.
12. Bothorel, P., Belle, J., Lemaire, B., *Chem. Phys. Lipids* (1974) **12**, 96.
13. Nagle, J. F., *J. Chem. Phys.* (1973) **58**, 252.
14. Marčelja, S., *Intern. Liquid Crystal 5th, Stockholm, 1974.*
15. Nagle, J. F., *Proc. Natl. Acad. Sci. U. S. A.* (1973) **70**, 1443.
16. Gaffney, B., McConnell, H. M., *J. Magn. Reson.* (1974) **16**, 1.
17. Gaffney, B., McConnell, H. M., *Proc. Natl. Acad. Sci. U.S.A.* (1971) **68**, 1274.

18. Poldy, F., Dvolaitsky, M., Taupin, C., *J. Phys. Paris Collog.* (1975) **36,** 1–27.
19. Charvolin, J., Rigny, P., *J. Chem. Phys.* (1973) **58,** 3999.
20. Bloom, M., *Specialized Colloque Ampère, Cracow, 1973.*
21. Roberts, R. T., *Nature London* (1973) **242,** 348.
22. Ågren, G., *J. Phys. Paris* (1972) **33,** 887.
23. Rouse, P. E., *J. Chem. Phys.* (1953) **21,** 1272.
24. Chapman, D., Salsbury, N. J., *Trans. Faraday Soc.* (1966) **62,** 2607.
25. Tiddy, G. J. T., "NMR of Liquid Crystals and Micellar Solutions," Chemical Society Annual Report on NMR, London, 1974.
26. Chapman, D., *Intern. Liquid Crystal Conf., 5th, Stockholm, 1974.*
27. Deloche, B., Charvolin, J., Liébert, L., Strzelecki, L., *J. Phys. Paris Colloq.* (1975) **36,** 21.
28. Levine, Y. K., Birdsall, N. J. M., Lee, A. G., Metcalfy, J. C., Partington, P., Roberts, G. C. K., *J. Chem. Phys.* (1974) **60,** 2890.
29. Brochard, F., de Gennes, P. G., in "Liquid Crystals," *Proc. Intern. Liquid Crystal Conf. 4th, Bangalore, 1973.*
30. Tardieu, A., Luzzati, V., Reman, F. C., *J. Mol. Biol.* (1973) **75,** 711.
31. Hervet, H., Volino, F., Dianoux, A. J., Lechner, R. E., *Phys. Rev. Lett.* (1975) **34,** 451.

RECEIVED November 19, 1974.

8

Biowater

E. FORSLIND

Division of Physical Chemistry, The Royal Institute of Technology, 100 44 Stockholm 70, Sweden

Starting from the mixture model, the structural behavior of water in the presence of dissolved simple ions is discussed from the point of view of defect formation and lattice distortions at interfaces. The observed behavior of the ions and the water lattice is applied to a number of unsolved biological problems in an attempt to elucidate the specific interface phenomena that are characteristic of such systems.

Although it is well recognized that the overwhelming majority of biological reactions take place in dilute aqueous solutions, very little attention has been paid to the water as a reaction partner, especially from the structural point of view. I shall discuss a few problems that are associated with the presence and the function of water in biological systems.

A Model for the Structure of Liquid Water

Starting from the mixture model of water, which is based on the introduction of lattice defects in the thermally expanded ice lattice, we shall first consider some ion–water interactions that modify the structure of water. An illustrative example is found when the solubilities of perchlorates are compared. In general, solubilities are high. However, potassium, rubidium, and cesium perchlorates have a very slow solubility in cold water; this, in fact, is utilized in the manufacture of $KClO_4$. The reduction in solubility can be traced back to the structural properties of water. Although the conformation of the ClO_4^- tetrahedron is similar to that of the water tetrahedron that it replaces, the difference in geometrical dimensions prevents a simple substitution. On dissolution of the ClO_4^- ion, a certain amount of water lattice distortion must occur in order to achieve a reasonable fit between the two structures.

As a rule, the distortion of the water lattice that is found in water without a solute (*1*) can easily take place in cooperation with the accompanying cation except in the cases of potassium, rubidium, and cesium. These ions are large enough to fill the cavities of the water lattice and to attenuate the lattice vibrations, thus preventing a local collapse of the structure and an increase in the number of interstitial water molecules. The normal water structure is essentially retained, and the lattice, stabilized by cations of the proper size, rejects the complex nonfitting ion (*2*).

The case of perchlorate behavior illustrates how an apparently simple, steric effect in one solute component can profoundly affect the interaction between the water and the whole solute. In an attempt to elucidate the solvation mechanism of simple electrolytes, Bergqvist and Forslind (*3*) made an NMR investigation of the water proton resonance in alkali halide solutions. They observed chemical shifts that were in unambiguous agreement with the findings of Hertz and Spalthoff (*4*) and Hindman (*5*), but they developed a different scheme for the separation of the shifts that are produced by cations and anions. By using the proton resonance shifts to measure the mean energy change of the hydrogen bonds, and the ionic volumes and the polarizabilities to characterize the solutes, they accounted for the electronically favored ionic charge distributions in the interaction with the water lattice. They assumed the same polarization proportionality factors for all cations and a common polarization mechanism for cations and anions, which is reasonable for ions with comparable electronic structures. Finally they neglected the coupling crossterms, *i.e.* the nonadditive contributions to the shift.

The assumptions were tested on the observed shifts of 19 alkali halide salts; the results were very satisfactory except in two cases, those of lithium and fluorine. Disregarding for the moment the behavior of the lithium and fluorine ions, the picture that emerges from the investigation may be summarized as follows.

There is a steric contribution to the main shift that is always positive, *i.e.* it corresponds to a decrease in hydrogen bond energy. In other words, the steric disturbance is always disruptive, displacing the oxygen atoms from their normal mean positions. It is counteracted by a polarization shift that is always negative, *i.e.* it corresponds to an increased hydrogen bond energy. The two effects practically cancel each other in the case of potassium, whereas the negative contributions dominate for the smaller and the larger cations as expected. However, the shift for the lithium ion, as determined heuristically, is far beyond the value expected from the polarizing shift contribution, which is expected to be very weak for lithium. This may indicate a considerable contribution to the negative shift that is derived from the vibrational attenuation of the water lattice attributable to interstitials. For all the anions except fluorine, the steric positive shift dominates over the negative polarizing contribution. The deviation in the fluorine shift, as determined heuristically, is expected for reasons already cited.

These and similar findings have led to the concept of structure makers and structure breakers among the ions (*6*). Thus, the alkali ions are all considered structure makers. Of the halide ions, Cl^-, Br^-, and I^- are all structure breakers, whereas F^- is an exception.

There is, however, a second way of defining the structure breakers that is based on the behavior of the water oxygens under the effect of the solutes (*7*). When the number of interstitial water oxygens exceeds the normal concentration (*1*), we obviously are dealing with a structure breaker. The process may be reversed, and by extending the definition in Ref. 7, we conclude that structure making occurs when the number of interstitial oxygens decreases below the normal value. This can occur, for example, at an interface that displays specific adsorption of water. The coupling of water molecules by hydrogen bonding to a heavier or more rigid substrate reduces the thermal vibrational amplitudes of the water lattice. The thermodynamic equilibrium of the lattice, which is normally maintained by the attenuating effect of the interstitial molecules which strike a balance between defect formation and lattice vibrations, can now be attained in part by the coupling to the substrate. The implication, of course, is a reduction in the number of interstitials needed to maintain equilibrium and a consequent decrease in water density. Both effects, strengthening of the hydrogen bonds and decrease in density, can occur simultaneously as is evidenced by the well known relative viscosity plots of the alkali halides *vs.* concentration (*8*).

Biological Processes

We may now assemble the foregoing information into a molecular description of a few biological processes in which the interaction between water and metal ions plays an important role. First some problems related to signal transfer in nerve cells are discussed. This is followed by some comments on the mechanism operating at nerve synapses in which, in addition to the sodium and potassium ions, a specific transmitter substance and calcium ions take part.

Sodium and potassium ions are vital to the normal functioning of the nerve cell. The ions are separated by the cell membrane with sodium on the outside and potassium on the inside of the resting cell. A model of the basic membrane, which in principle was built as a bilayer according to the well known Davson–Danielli–Robertson scheme but which included water as a structural component, was presented in a short communication at this conference. The model is described in some detail in two recent publications (*9, 10*). Here I shall be content to summarize the principal features of the lecithin–water system with emphasis on the water structure that is assumed to determine the ionic transport through the membrane.

The positively charged choline end group of the lecithin molecule is fitted into a water lattice in such a way that the lecithin molecule takes up an area of about 57 A^2 projected on the membrane surface. The water molecules belonging to the water lattice become coupled to the hydrophilic oxygens of the phosphate groups and ester linkages, thereby forming a continuous network of water bridges. The quarternary nitrogens, each surrounded by three methyl groups, substitute a water molecule; the methyls become embedded in the adjacent cavities of the water lattice, avoiding interference with the water structures. The lone pair electrons of the immediately adjacent water molecules are directed toward the positive nitrogen, and the strongly hydrophobic, positively charged group is thus naturally accommodated by the water lattice.

Let us now consider the ionic transport through the bilayer. The spacing between the alkyl chains of the lecithins is large, some 7 A. Diffusional processes can obviously take place as in a water lattice, and a positive charge is needed to keep the cations in place on either side of the membrane, permitting a limited transfer of water and anions. The lecithin molecule can be *either* (a) a neutral dipole with a positive charge at the quarternary nitrogen and a negative charge on the phosphate group which has ejected a proton to the water lattice, or (b) positively charged with the proton returned to the phosphate group. Let us consider the latter state as normal; the membrane is at rest, and it forms a barrier to the cations. Then all that is needed to open the barrier is removal of the protons from the phosphate groups. A possible mechanism to produce this effect is provided by proton transfer in the water bridges that link the phosphate groups in the essentially two-dimensional interlecithin network that forms part of the surrounding water lattice. We shall return to this trigger mechanism later.

First, however, let us see what can happen to the external and internal positive ions. The sodium concentration is higher outside the membrane while the opposite is true of the potassium ion. Furthermore, the potassium ion fits well into the cavity of the water lattice, and the sidium ion causes the cavity to contract without, however, producing a local collapse and rupture of the hydrogen bonds to the remainder of the lattice.

Inside the cell membrane, the axoplasm forms a gel, moving slowly away from the centers of protein synthesis in the main cell body. The axon appears to contain specific channels, the microtubuli, for the more rapid transport of metabolites; these can also be expected to permit fairly rapid movement of cations along the axon. Inside the gel surrounding the microtubuli there is water that is structurally stabilized by hydrogen bonding to the solid gel phase and by the formation of internal interfaces around hydrophobic groups of the solid. It likewise seems reasonable to assume that a very important stabilization mechanism is in operation because of the fit of potassium ions into the water cavities. There are several known cases of such stabilization in other gel-forming systems, *e.g.* in clay gels either in their native state or in drilling muds. Let us therefore assume that the axoplasm is stabilized by potassium, which is obviously the ion that is energetically preferred to the gel. One other possible reason for the specific adsorption of the cation is the tendency of every hydrogel to exchange protons with the water phase; this is compensated for by counterions with a long lifetime in the adsorbed state.

Having thus advanced some arguments for the selective fixation of potassium within the gel, we note that the local ion transport process at the Ranvier node, consequent to a nerve impulse, is always initiated by the entrance of sodium followed by the expulsion of potassium. It appears that the condition for potassium release is not, at first, the trigger process but the entrance of sodium into the axoplasm. The possible sequence of events may then be as follows.

Removal of the phosphate protons from the lecithins by the trigger process turns a few of the positive choline end groups down toward the negative phosphate oxygen which permits a few sodium ions to penetrate the water layer of the membrane. The water lattice distortion around a sodium ion in bulk water is enhanced in the membrane because of the disturbance around the choline methyls that occurs on their reversal, and an avalanche of sodium ions is released in the opening. The sodium ions, penetrating into the axoplasmatic gel, disturb the water structure by competing with the potassium ions that stabilize the water and act as counterions to the negative charge left on the solid phase as the result of protolysis. A certain number of potassium ions are released, and they move out of the cell along the concentration gradient. As they enter the disturbed membrane, they begin to compensate the remaining negative charges on the lecithin; they repel the positive choline end groups to their normal positions and rapidly block further entrance of sodium ions. The potassium ion depletion is eliminated by ions that move, slightly delayed, in the microtubuli to the zone of disturbance. Order is restored.

The disturbance has, however, generated a new proton wave (*2*) that is propagated as a back shift of protons along the hydrogen bond network to the next Ranvier node where the process is repeated. It is an open question how much the myelin sheaths participate in this pulse transfer. They are certainly built for the task, and the morphology of the junction between the cell membrane and the Schwann cell would favor a close cooperation.

The Synaptic Mechanism

An interesting difference is found between the frequency-modulated signal transfer in nerve cells and the amplitude-modulated mechanism operating at synapses in which specific transmitter substances and calcium ions are involved as well as sodium and potassium. Calcium and sodium ions have about the same Pauling radius and comparable ionic volumes. With calcium, however, the charge is doubled and so is the ionic field. This leads to a water lattice distortion in which the hydration shell of six octahedrally coordinated water molecules is separated from the remainder of the lattice, essentially as the result of unfavorable orientation of the hydrogen bonds. The distortion is not as pronounced as it is with magnesium ion which has the same charge as calcium but a smaller ionic diameter. In fact, the volume of the magnesium–water hydration complex is less than 50% that of the corresponding coordination complex in liquid water.

I shall assume that we are concerned with a cholinergic transmitter. Close to the presynaptic membrane, clusters of small vesicles are heaped ready to empty their contents of acetylcholine into the synaptic cleft. A certain amount of spontaneous acetylcholine ejection does occur, however, and this gives rise to a random distribution of small signals when the transmitter substance reaches the postsynaptic membrane. Unless there is a coincidence in space and time of the spontaneous ejection events to make several signals cooperate, the barrier to a propagating impulse in the postsynaptic membrane is never exceeded. The ejection mechanism is apparently unknown although it is obvious that the arrival of the motor nerve impulse is somehow connected with the release of the vesicles. The synaptic cleft is some 50–100 A wide; moreover, it abounds in acetylcholine esterase which rapidly breaks down the molecules to make them ineffective. It is evident that a substantial release of transmitter substance is needed in order to produce a propagating signal.

Now, it has been observed that calcium ions are needed (*11, 12, 13, 14*) for the release of acetylcholine. If we assume the same trigger mechanism as was described previously, we see that the penetration of calcium gives rise to a much stronger water lattice and membrane structure breakdown than does that of sodium. The bilayer membrane of the vesicles, which is tightly coupled to the presynaptic membrane *via* the cytoplasmatic water, is affected equally strongly and the probability of vesicle breakdown is enhanced. The released acetylcholine gives rise to the characteristic two-phase signal in the sarcolemma and its invaginations into the transverse tubule system. There the propagated pulse similarly releases the calcium ion avalanche from the terminal cisternae into the sarcoplasmic reticulum and the myofibrils, thereby initiating the muscle contraction.

Literature Cited

1. Forslind, E., *Q. Rev. Biophys.* (1971) **4** (4), 325.
2. Forslind, E., Jacobsson, A., "Water: A Comprehensive Treatise," F. Franks, Ed., Vol. 5, Plenum, New York, 1974.
3. Bergqvist, M., Forslind, E., *Acta Chem. Scand.* (1962) **16,** 2069.
4. Hertz, H. G., Spalthoff, W., *Z. Elektrochem.* (1959) **63,** 1096.
5. Hindman, J. C., *J. Chem. Phys.* (1962) **36,** 1000.
6. Hertz, H. G., Zeidler, M. D., *Ber. Bunsenges.* (1970) **69,** 821.
7. Walrafen, G. E., *J. Chem. Phys.* (1970) **52,** 4176.
8. Bernal, J. D., Fowler, P. H., *J. Chem. Phys.* (1933) **1,** 515.
9. Forslind, E., Kjellander, R., *Sven. Naturvetensk.* (1974) **9,** 174.
10. Forslind, E., Kjellander, R., *J. Theor. Biol.* (1975) **51,** 97.
11. Weber, A., "Bioenergetics," D. R. Sanadi, Ed., Vol. 1, p. 203, Academic, New York, 1966.
12. Ottoson, D., "Nervsystemets Fysiologi," Natur och Kulturs, New York, 1970.
13. Karlson, P., "Biochemie," Georg Thieme, Stuttgart, 1967.
14. Lehinger, A. L., "Biochemistry," Worth, Stockholm, 1970.

RECEIVED November 19, 1974.

9

Ion Binding and Water Orientation in Lipid Model Membrane Systems Studied by NMR

GÖRAN LINDBLOM and NILS-OLA PERSSON

Division of Physical Chemistry 2, The Lund Institute of Technology, Chemical Center, Lund, Sweden

GÖSTA ARVIDSON

Department of Physiological Chemistry, University of Lund, Lund, Sweden

The static quadrupole effects of alkali metal ions and deuterons in NMR spectra of anisotropic mesophases are discussed. Lamellar lyotropic liquid crystals composed of lecithin, heavy water, and varying concentrations of sodium chloride and cholesterol were studied at different temperatures. The observed quadrupole splittings of ^{23}Na and ^{2}H in these systems are interpreted to indicate that the number of water and sodium binding sites increases with increasing temperature or electrolyte concentration. A model is proposed where the increased binding of sodium ions and water is related to a conformational change of the phosphorylcholine group. The results obtained on cholesterol-containing samples may indicate that a rearrangement of the molecular organization of the bilayer occurs at a molar ratio of 4:1 between lecithin and cholesterol.

Several recent investigations using various physicochemical methods have provided convincing evidence to support the contention that the basic structure of most biological membranes consists of a phospholipid bilayer (*1, 2, 3, 4*). Studies on phospholipid model membranes can therefore be expected to yield relevant information on the role played by phospholipids in determining the characteristic properties of biological membranes (*5*). One important aspect of this problem concerns the mechanisms of interaction between the phospholipids and other membrane constituents such as cholesterol, proteins, and different inorganic ions. The importance of the hydrophobic parts of the phospholipid molecules in determining the general properties of the interior of the phospholipid bilayer has been investigated in great detail since the pioneering work of Chapman *et al.* (*6*). Far less is known about the ionic interactions involving the polar head-groups at the surface of the phospholipid bilayer. Obviously much more work on model membrane systems is needed before we can acquire a detailed understanding at the molecular level of physiologically important, ionic processes in biological membranes—*e.g.*, the selective transport of alkali metal ions, the interaction with divalent cations such as Ca^{2+} and Mg^{2+}, ionic interactions with proteins, and the activation of certain membrane-bound enzymes by metal ions. Among the problems to be elucidated are the conformation and orientation of the polar head groups of the phospholipid molecules, the structure of water at the bilayer surface, and the interrelationships between the ionic structure of the polar head groups and the molecular organization of the hydrophobic part of the bilayer. The aim of the present work is to demonstrate some of the potentialities of NMR spectroscopy in investigations of ionic interactions and water structure in lamellar mesophases of lipids. Lamellar (or smectic) mesophases of amphiphilic lipids are composed of lipid bilayers intercalated by water spaces (*7*). Extensive work during recent years has proved that lamellar mesophases provide excellent model systems for membrane research (*8*).

Quadrupole Splittings in NMR Spectra

Static quadrupole effects in NMR are observed in solids (*9*) and also in anisotropic liquid crystals (*10, 11, 12*). For nuclei with spin quantum numbers, *I*, greater than ½, the distribution of positive charge over the nucleus can be nonspherical and the situation can be described in terms of a nuclear electric quadrupole moment. The interaction between the quadrupole moment, *e*Q and electric field gradients, *e*q, shifts the energy levels of the nuclear spin states.

The nuclear spin hamiltonian (H) for the Zeeman (H_Z) and the quadrupole (H_Q) interactions may be written

$$H = H_Z + H_Q = -\nu_L I_Z + \beta_Q \sum_{q=-2}^{2} (-1)^q V_{-q} A_q \qquad (1)$$

H_Q is expressed in frequency and $\beta_Q = \dfrac{eQ}{2I(2I-1)h}$. ν_L is the Larmor precession frequency. V_q values represent the irreducible components of the electric field gradient tensor, and A_q values are irreducible tensor operators working on the nuclear spin functions. All these tensors are of the second rank. V_q and A_q in Equation 1 can be expressed in any coordinate system. It is most convenient, however, to have the spin operators given in a laboratory-fixed coordinate system since the nuclear spin is quantized along the applied magnetic field. The electric field gradients, on the other hand, are most conveniently expressed in a principal axis coordinate system fixed at the nucleus. The quadrupolar hamiltonian can then be written:

$$H_Q = \beta_Q \sum_{qq'} (-1)^q V_{-q}{}^M A_{q'}{}^L D_{q'q} (\Omega_{LM}) \tag{2}$$

where $D_{q'q}$ is a Wigner rotation matrix element of the second rank, and Ω_{LM} is shorthand notation for the appropriate Euler angles specifying the transformation from the molecular frame (M) to the laboratory frame (L). If the molecular motion has to be taken into account, the hamiltonian in Equation 2 becomes time dependent through the time dependence of Ω_{LM}. When the fluid is isotropic, in the sense that all orientations are equally probable, the mean values of the second-order Wigner rotation matrix elements are zero, and the quadrupole interaction contributes only to relaxation. In an anisotropic medium like a lamellar or hexagonal liquid crystal the mean value of H_Q is no longer zero, and one can observe quadrupole splittings in the NMR spectrum.

Consider a lamellar mesophase, being macroscopically aligned so that the symmetry axis, referred to as the director, has the same direction throughout the sample. If the transformation from the molecular coordinate system to the laboratory system is performed *via* the director coordinate system (D), Equation 2 reads

$$H_Q = \beta_Q \sum_{qq'q''} (-1)^q V_{-q}{}^M A_{q''}{}^L D_{q'q} (\Omega_{DM}) D_{q''q'} (\Omega_{LD}) \tag{3}$$

Here Ω_{ij} specify the transformation from coordinate system j to system i. In Equation 3 only $D_{q'q}$ (Ω_{DM}) varies with the molecular motion. Since amphiphilic liquid crystalline systems generally are cylindrically symmetrical around the director $\overline{D_{q'q} (\Omega_{DM})} = 0$ if $q' \neq 0$. If it also is assumed that a nucleus stays within a domain of a given orientation of the director over a time that is long compared with the inverse of the quadrupole interaction, one now obtains for the static quadrupole hamiltonian

$$\overline{H}_Q = \beta_Q \sum_{qq''} (-1)^q V_{-q}{}^M A_{q''}{}^L \overline{D_{oq} (\Omega_{DM})} D_{q''o} (\Omega_{LD}) \tag{4}$$

In the limit where the nuclear Zeeman term in the nuclear spin hamiltonian is much larger than the quadrupole interaction, it is only the secular part of H_Q that contributes to the time-independent hamiltonian, H_o.

$$H_o = -\nu_L I_z + \frac{\nu_Q S}{6} (3 \cos^2 \theta_{LD} - 1) (3I_z^2 - I^2) \quad (5)$$

where $\nu_Q = \frac{3}{2} \beta_Q eq$ and the order parameter

$$S = \frac{1}{2} \{\overline{(3 \cos^2 \theta_{DM} - 1)} + \eta \overline{\sin^2 \theta_{DM} \cos 2\phi_{DM}}\}$$

Here θ_{DM} is the angle between the director and the largest component, eq, of the electric field gradient tensor, and η is the asymmetry parameter. The different coordinate systems used are shown in Figure 1.

The hamiltonian in Equation 5 gives, for the case where the spin quantum number is equal to 3/2 (as for ^{23}Na) four energy levels (*cf.* Figure 2).

$$\begin{aligned} E_{3/2} &= -3/2\ \nu_L + 1/2\ \nu_Q S\ (3 \cos^2 \theta_{LD} - 1) \\ E_{1/2} &= -1/2\ \nu_L - 1/2\ \nu_Q S\ (3 \cos^2 \theta_{LD} - 1) \\ E_{-1/2} &= 1/2\ \nu_L - 1/2\ \nu_Q S\ (3 \cos^2 \theta_{LD} - 1) \\ E_{-3/2} &= 3/2\ \nu_L + 1/2\ \nu_Q S\ (3 \cos^2 \theta_{LD} - 1) \end{aligned} \quad (6)$$

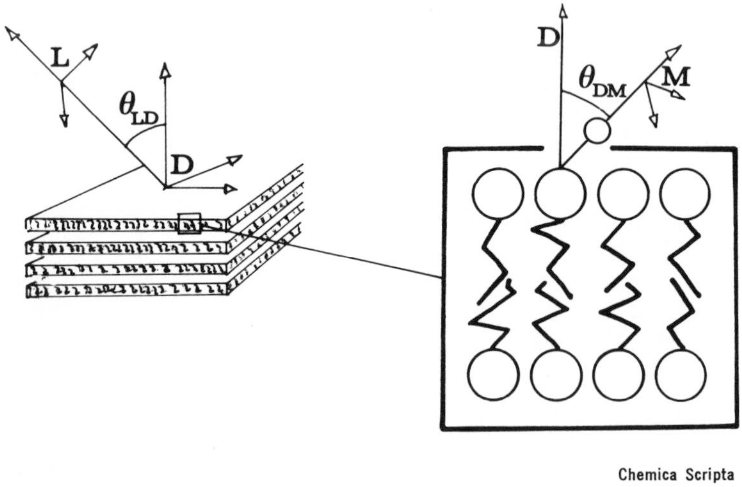

Chemica Scripta

Figure 1. Schematic of the mesomorphous structure in a lamellar phase. The different coordinate systems used in the text are outlined: laboratory frame (L), director frame (D), and molecular frame (M). θ_{LD} and θ_{DM} are angles between the z axis in laboratory–director systems and director–molecular systems, respectively (13).

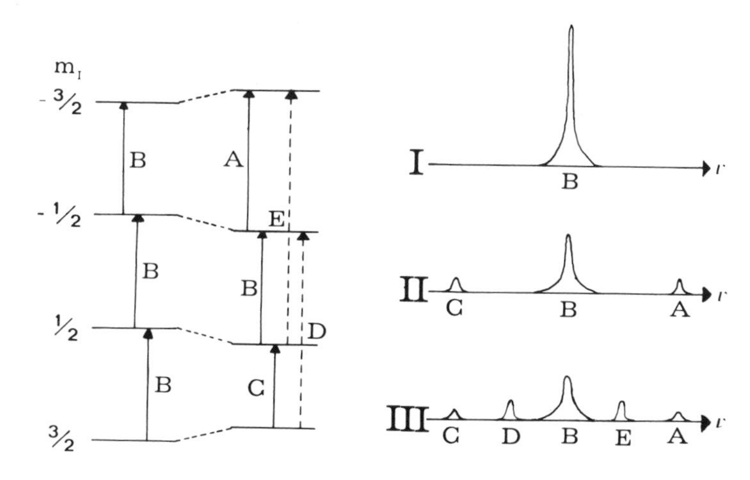

Figure 2. Energy levels for $I = 3/2$ *with* $m_I = \pm 1/2$ *and* $m_I = \pm 3/2$ *showing NMR transitions without (B) and with (A and C) a first-order quadrupole splitting and with double quantum transitions (D and E). The "allowed" transitions* $\Delta m_I = \pm 1$ *and the* $\Delta m_I = \pm 2$ *transitions are indicated. The corresponding spectra are shown to the right: (I) spectrum for* $\nu_Q S\,(3\cos^2\theta_{LD} - 1) = 0$*; (II) spectrum to first order in quadrupole interaction; and (III) spectrum showing transitions where* $\Delta m_I = \pm 2$.

For deuterons the spin quantum number is equal to 1, and Equation 5 gives three energy levels.

For an oriented sample with a uniform director orientation the ^{23}Na NMR spectrum at a weak radiofrequency field thus consists of three equally spaced peaks. The relative magnitude of the three resonance lines is governed by the transition probabilities between the different energy levels. It turns out (*9*) that the central line contributes 40% of the total intensity while each of the satellites contributes 30%. The deuteron NMR spectrum consists of two equally intense peaks.

We define an experimental quantity, which we call the quadrupole splitting (*13*), as the distance (in Hz) between two adjacent peaks obtained in the CW (continuous wave) NMR spectrum, corresponding to the distance A–B or B–C in Figure 2. For an oriented liquid crystalline sample one obtains from Equation 6:

$$\Delta(\theta) = |\,\nu_Q S\,(3\cos^2\theta_{LD} - 1)\,| \tag{7}$$

Comparison of experimental data with Equation 7 makes it possible to determine how the director is oriented with respect to the constraint (*14*) responsible for macroscopic alignment.

When a polycrystalline material or an unaligned anisotropic mesophase is used, the spectral shape of the NMR signal shows a typical pow-

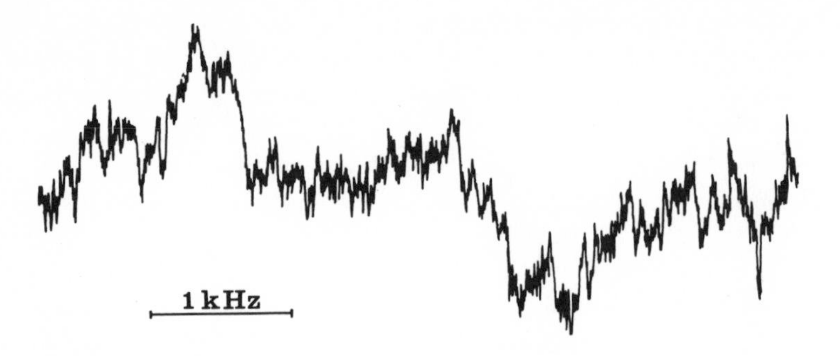

Figure 3. Continuous wave ^2H NMR spectrum of a lamellar liquid crystalline sample containing dimyristoyllecithin with a choline–N–C^2H_3 group. The water content of the sample was 25% (w/w), and the temperature was 25°C.

der pattern. This spectral shape is obtained since here all values of $\cos \theta_{LD}$ are equally probable and the distance between major absorption maxima in the NMR spectrum, Δp, corresponds to that for $\theta_{LD} = 90°$ of a marcoscopically aligned sample. The symbol Δp for a "powder" sample is given by

$$\Delta p = | \nu_Q S | \tag{8}$$

A typical powder spectrum of ^2H for a C^2H_3 group located at the choline nitrogen of a phospholipid molecule, is shown in Figure 3. Equations 7 and 8 are valid if it is sufficient to consider only one binding site for the deuteron or ionic nuclei studied. In the systems considered, more than one type of binding site exists, and an exchange of species is occurring between the sites. In cases where this exchange is much faster than the splitting difference only one splitting in the NMR spectrum is observed, and this is given by

$$\Delta p = | \Sigma\, p_i\, \nu_Q{}^i\, S_i | \tag{9}$$

where p_i is the fraction of deuterons or ions in site i. In some lyotropic liquid crystalline systems containing heavy water and a compound having an $-OH$ (*15*) or an $-NH_2$ (*16*) group, for example, the deuterons can exchange between the water and the alcohol or amine molecules. Recently it was shown (*15*) experimentally that slow or intermediately fast deuteron exchange markedly changes the spectral shape. In the region of intermediate exchange rate a broadening of the NMR signal appears, and from NMR powder spectra exhibiting such broadening caused by exchange, it was possible to determine the exchange time (*15*). It is interesting that this method makes it possible to determine exchange times much shorter than those which can be obtained using ordinary proton NMR methods.

In systems where the deuterons studied are in the limit of slow exchange, information about the order parameter often can be extracted. For deuterons it is a good approximation to assume that the quadrupole coupling constant, being of intramolecular origin, is independent of sample composition and temperature. From experimental data we can then use Equations 7 or 8 to calculate the absolute value of the order parameter from an assumed value of the quadrupole coupling constant. This method has been used for studies of the degree of orientation of the aliphatic chains of soap molecules (*17, 18*) and of hydroxyl group deuterons of alcohols of lamellar mesophases (*19*). For ions in solution the situation is complicated by the absence of any knowledge of the quadrupole coupling constant. Therefore, it is difficult to extract information on the order parameter from ^{23}Na quadrupole splittings. However, by using a simple electrostatic model it is possible to rationalize the counterion quadrupole splittings for at least some amphiphilic systems (*13*).

Ref. *13* contains a more complete discussion of static quadrupolar effects for amphiphilic mesophases. That work also includes a treatment of counterion quadrupole relaxation for liquid crystalline systems; a brief outline of this discussion is given in the next section.

Quadrupolar Relaxation

For nuclei which possess electric quadrupole moments, a generally very powerful relaxation mechanism is afforded by the interactions of the quadrupole moment with fluctuating electric field gradients. For such nuclei, this time-dependent interaction is usually the dominant relaxation mechanism. No well-defined relaxation times (T_1 or T_2) exist for nuclei with spin quantum numbers, $I > 1$ unless the molecular motion is rapid with respect to the resonance frequency at which the measurement is made. For a nucleus having $I \geqslant 3/2$ the relaxation times are well defined (*20*) only in the limit of "extreme narrowing"—*i.e.*, $J(\omega_o) = J(0)$, where J is a spectral density as defined below and ω_o is the Larmor precession frequency (this condition is the same as saying that $\omega_o\tau_c << 1$ where τ_c is the correlation time characterizing the time dependence of the interaction).

$$1/T_1 = 1/T_2 = 24\pi^2(2I-1)(2I+3)\ J_{oo}(0) \tag{10}$$

where

$$J_{q'q}(\omega) = \frac{1}{2}\ \beta_Q{}^2 \int_{-\infty}^{+\infty} \overline{V_q{}^L\ (0)\ V_{q'}{}^L\ (\tau)}\ e^{-i\omega\tau} d\tau$$

The NMR spectral shape is a Lorentzian curve, and the linewidth, $\Delta\nu_{1/2}$, at half-height of the absorption peak, gives a measure of the spin relaxation rates through $\Delta\nu_{1/2} = 1/\pi T_2$.

When $J(0) \neq J(\omega_o)$ and $I = 3/2$, the NMR signal consists of two superimposed Lorentzian curves (a and b) with the relative intensities of 3:2 and linewidths given by (*13, 20*):

$$\Delta\nu^a{}_{1/2} = (16\pi/5)\ \nu_Q{}^2(1+\eta^2/3)\{\tilde{J}_{oo}(0) + \tilde{J}_{-11}(\omega_o)\} \tag{11a}$$

$$\Delta\nu^b{}_{1/2} = (16\pi/5)\ \nu_Q{}^2(1+\eta^2/3)\{\tilde{J}_{-11}(\omega_o) + \tilde{J}_{-22}(2\omega_o)\} \tag{11b}$$

where $\tilde{J}$ is a reduced spectral density defined by

$$\tilde{J}_{q'q}(\omega) = \beta_Q{}^{-2}\ [\overline{V_q{}^L(0)\ V_{q'}{}^L(0)}]^{-1}\ J_{q'q}(\omega)$$

Recent experimental observations have shown that a situation with two superimposed lorentzian curves in the NMR spectra for sodium ions occur in some biological tissues (*21, 22*).

The expressions for the linewidths in a spectrum with quadrupole splittings for nuclei with $I = 3/2$ has been derived (*13*). Thus the central peak has the same linewidth as the narrow component in Equation 11b. In Ref. *13* the linewidths of the two satellites have also been calculated to be:

$$\Delta\nu_{1/2} = (16\pi/5)\nu_Q{}^2(1+\eta^2/3)\{\tilde{J}_{oo}(0) + \tilde{J}_{-11}(\omega_o) + \tilde{J}_{-22}(2\omega_o)\} \tag{12}$$

In the systems under consideration the ions exchange between different binding sites rapidly compared with the rate of relaxation, and the observed linewidth is given by:

$$\Delta\nu_{obs} = \Sigma\ p_i\Delta\nu_i \tag{13}$$

where p_i is the fraction of ions in site i characterized by the intrinsic linewidth $\Delta\nu_i$. The effect of exchange on the relaxation rate of quadruple nuclei is discussed in more detail in Ref. *13*. Thus there are three main factors that determine the observed linewidths—namely, the electric field gradients, the spectral densities, and the distribution of the ions over the different binding sites.

Material and Methods

Dimyristoyl- and dipalmitoyllecithin were synthesized by acylation of sn-glycero-3-phosphorylcholine with fatty acid anhydrides as described by Cubero Robles and van den Berg (*23*). The sn-glycero-3-phosphorylcholine was prepared by deacylation of egg-yolk lecithin according to

Brockerhoff and Yurkowski (*24*), and the fatty acid anhydrides were synthesized according to Selinger and Lapidot (*25*). Phosphatidylethanolamine was prepared from freshly extracted egg-yolk lipids by column chromatography on alumina and silicic acid. [*Me*-^{2}H]Dimyristoyllecithin was synthesized by methylation of dimyristoylphosphatidyl-*N,N*-dimethylethanolamine with [^{2}H]methyl iodide (from Geigy) in the presence of cyclohexylamine as described by Slotboom *et al.* (*26*). Dimyristoylphosphatidyl-*N,N*-dimethylethanolamine was obtained by demethylation of dimyristoyllecithin with sodium benzenethiolate according to Stoffel *et al.* (*27*). All lecithin preparations were purified by silicic acid column chromatography. Cholesterol (Merck AG) was recrystallized from ethanol several times before use. The purity of all lipid preparations was checked by thin-layer chromatography on silica gel in several different solvent systems.

In magnetic resonance one measures a component of the magnetization, $M(t)$. This is usually obtained in one of two ways: (1) by "preparing" the spin system in a given nonequilibrium initial state through the application of a radiofrequency (rf) field pulse, or (2) by pertubing the system under study continuously by applying a continuous rf wave. In the pulsed NMR experiment the decay of the magnetization is directly observed as a function of time, whereas in the continuous wave (CW) NMR experiment the magnetic resonance spectrum is observed, and this spectrum can be associated with the Fourier transform of the time-dependent magnetization.

When one wishes to measure, for example, large quadrupole splittings or broad linewidths, a wideline NMR instrument is generally used. In this work the ^{23}Na and ^{2}H CW NMR measurements were done with a Varian V-4200 NMR spectrometer, which is a broadline instrument. The recording of experimental data has been described previously (*28, 29, 30*). For samples where the intensity is not sufficient for a single CW NMR spectrum and when the splitting is not too large and the relaxation time is short enough, it is advantageous to use pulsed NMR. The perturbation of an intense rf pulse produces a magnetization which has a rotating component in the plane perpendicular to the external magnetic field. The rotating transverse magnetization induces an electric current in a coil enclosing the sample, and the observable pulsed NMR signal is described by this transverse magnetization. The specific pulse used in the experiment is identified by the number of angular degrees through which it rotates the nuclear spin magnetization. The expectation value of the transverse magnetization following a 90° pulse is given by (*31*):

$$\langle M_+(t) \rangle = iM_o \exp(-i\omega_o t)\cos(\omega' t) \qquad (14)$$

where M_o is the equilibrium magnetization parallel to the applied magnetic field. The transverse magnetization rotates at the resonance frequency ω_o. A quadrupole splitting gives rise to the low frequency cosine

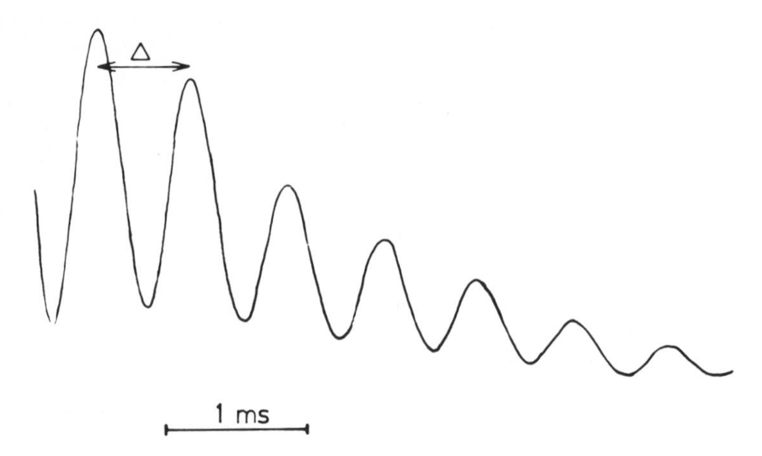

Figure 4. Deuteron resonance decay curve after a 90° pulse for the same sample as in Figure 3. The symbol △ represents the measured quadrupole splitting.

terms which modulate the amplitude of this high frequency rotating magnetization. Thus ω' in Equation 14 contains the same information as $\Delta(\theta)$ or Δp given in the above section. The pulsed NMR work was performed with a Bruker 322s pulse spectrometer. A typical quadrupole split pulsed NMR decay curve for the deuterated *N*-methyl group of lecithin in a lamellar mesophase is shown in Figure 4, where the quadrupole splitting, Δ (in ms), is also indicated. The great advantage of the pulse NMR experimentation is that the time required to obtain the NMR spectrum is several hundredths that required by CW methods. Furthermore, the sensitivity can be increased several times by time averaging several scans on a computer of averaged transients (CAT). This technique can give 10–100 fold enhancements in signal-to-noise ratios compared with CW methods. For the spectrum in Figure 4, 200 scans were used. A different pulsed NMR method for determining doublet splittings, based on relaxation time measurements (T_2 and T_3) has been developed by Woessner (*32*) (T_3 is obtained from 90°–90° pulse sequences).

Double Quantum Transitions

The appearance of a sharp central peak in the deuteron NMR spectrum for unaligned lyotropic liquid crystalline samples has been observed by several authors (*30, 33, 34*). This has been interpreted in terms of phase inhomogenities (*35*) or isotropic motion (*36*). However, recently Wennerström *et al.* showed for a lamellar amphiphile–water mesophase that this could be referred to double quantum transitions (*37*). It is expected that double quantum transition peaks appear in NMR spectra also for other quadrupolar nuclei in unoriented mesophases. It follows from the calculated energy levels for $I = 3/2$ given above that double quantum transitions may, at sufficiently strong rf fields, appear at the frequencies:

$$\omega_D = \frac{E_{-1/2} - E_{3/2}}{2} = \nu_L - 1/2\ \nu_Q\ S\ (3\ \cos^2\ \theta_{LD} - 1)$$

and

$$\omega_E = \frac{E_{-3/2} - E_{1/2}}{2} = \nu_L + 1/2\ \nu_Q\ S\ (3\ \cos^2\ \theta_{LD} - 1)$$

For a mesophase composed of sodium octanoate, decanol, and heavy water such double quantum transitions for ^{23}Na are easily observed, as shown in Figure 5. This figure shows that at low rf field amplitudes a typical ^{23}Na powder spectrum is obtained, but at high rf field strengths new peaks arise in the spectrum.

To obtain a good signal-to-noise ratio, it is often useful to work with a high rf field amplitude when measuring quadrupole splittings. Because it is easy to detect double quantum transitions for unaligned mesophases, one has to be careful if large rf fields are used. Whether an observed peak is attributable to double quantum transitions or not can be investigated from the saturation behavior (*cf.* Figure 5). However by using the pulsed NMR technique, the difficulties with double quantum transi-

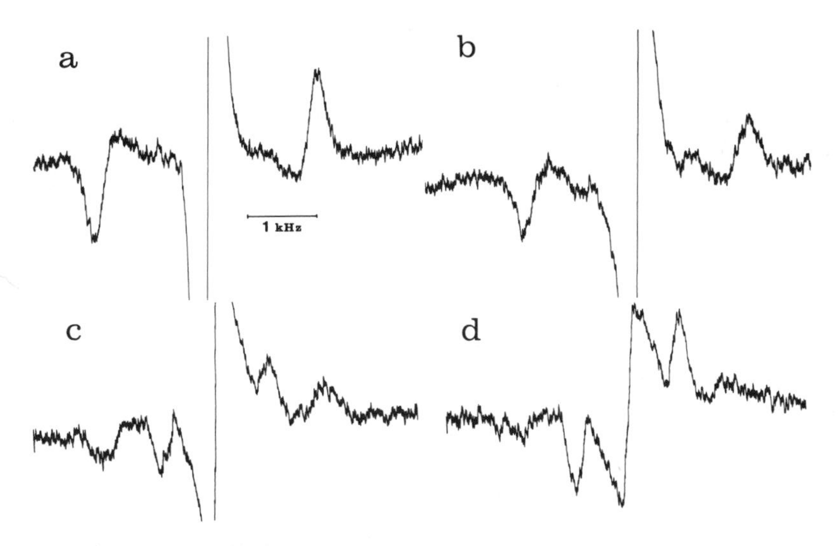

Figure 5. ^{23}Na NMR spectra of a lamellar mesophase sample of the system sodium octanoate–decanol–water. The rf field amplitude was increased in the order a,b,c,d. Peaks arising from double quantum transitions are clearly shown (cf. *Figure 1).*

tions are completely avoided since in the pulsed NMR experiment the rf field amplitude is zero when the spectrum is recorded.

Results and Discussion

^{23}Na NMR Studies. Various techniques have been used in studies of the role played by ions in different biological systems (*5, 38, 39*). In particular NMR methods have appeared to offer considerable promise. For some years we have been studying the binding of small ions and water molecules to colloidal aggregates. The NMR investigations were concerned with counterion linewidths (*40*), counterion quadrupole splittings (*29*), counterion chemical shifts (*41*), and heavy water deuteron quadrupole splittings (*19*). Note that all the alkali ions can be studied by NMR. Among the ions of particular interest in connection with the function of biological membranes (*e.g.*, Na^+, K^+, Ca^{2+}, and Mg^{2+}), sodium ions are best suited for NMR studies. Therefore our studies on the interactions of small ions in model membrane mesophases have been performed mainly with ^{23}Na NMR. Information has been obtained from investigations on both NMR linewidth and quadrupole splittings.

Our previous studies of ion binding in micellar solutions and lyotropic mesophases have shown (*42, 43*) that a change in the NMR linewidth can be interpreted in terms of a change in the interaction between the counterions and the charges at the amphiphilic surfaces. These investigations were done with systems containing an ionic amphiphile. Lecithins are zwitterionic molecules without net charge over a wide pH range. However, it is reasonable to assume that there is an interaction between sodium ions (from salt added to the samples) and the charged phosphate groups of the lecithin molecules since both ^{23}Na linewidths and ^{23}Na quadrupole splittings observed in these systems are quite large. We thus assume that the ion–ion interaction gives the dominant contribution to the electric field gradient determining the magnitude of linewidths and splittings obtained (*cf.* Ref. *13*). However, recently we observed ^{23}Na quadrupole splittings also for a nonionic amphiphilic liquid crystalline system (*44*), indicating that for at least some systems ion–dipole interactions must be taken into account.

In all our linewidth studies the width of the central peak was measured—*i.e.*, the linewidth obeying Equation 11b. In Ref. *45* we reported ^{23}Na linewidth studies on a lamellar mesophase containing egg-yolk lecithin. The effect of cholesterol on the linewidth was investigated, and we found that with increasing cholesterol content in the phospholipid bilayers a marked reduction in the linewidth was observed. In accordance with similar investigations for simple soap–alcohol–water lamellar mesophases (*42, 43*) this is interpreted as a partial release of sodium ions from the charged surfaces of the lamellae on addition of cholesterol. ^{23}Na linewidth studies of the effect of different anions on the sodium ion binding and of the competition between Na^+ and other alkali ions and also Ca^{2+} and UO_2^{2+} have been performed (*45*). It was found that neither Br^- or I^- nor Ca^{2+} or UO_2^{2+} had any significant effect on the sodium ion interaction. The competition experiment gave the following information about the relative affinities for the interaction of different ions with the lecithin–cholesterol lamellae: K^+ ions interact more strongly with the amphiphilic molecules than the Na^+ ions. The other alkali ions are bound more weakly than sodium ions, the affinities being in the sequence $K^+ > Na^+ > Rb^+ > Cs^+ \simeq Li^+$. Part of this affinity sequence was also obtained in similar experiments but with the aid of the quadrupole splitting measurements (*29*). Competition studies with sodium and calcium ions utilizing quadrupole splittings have also been performed with lamellar mesophases of egg yolk phosphatidyl ethanolamine and heavy water. When part of the sodium chloride was substituted by calcium chloride, the sodium quadrupole splitting decreased markedly, indicating, contrary to lecithin mesophases, a strong binding of Ca^{2+} to phosphatidylethanolamine molecules in the bilayer.

Sodium and deuteron quadrupole splittings in mesophase samples containing fully saturated, synthetic lecithins have also been investigated (*46*). The dependence of the sodium quadrupole splitting was studied with respect to (a) the cholesterol content of the phospholipid mesophase, (b) temperature, and (c) the sodium chloride concentration in the water layers. The results can be interpreted according to a simple physical model with only two binding sites—*i.e.*, the ions are either bound to the lamellae or are moving freely in the water layer. It is also assumed that ion exchange between these different sites is rapid compared with the difference in splitting between the sites.

Effect of Cholesterol on ^{23}Na Splitting. Figure 6 shows how the sodium quadrupole splitting depends on cholesterol content in dimyristoyl lecithin lamellae at four different temperatures. As may be inferred from this figure a minimum in the splitting, for all the temperatures investigated, is obtained at about 15 wt % cholesterol in the lipid bilayers, corresponding to a molar ratio of lecithin:cholesterol of about 4:1. The interpretation of the variation of the splitting has been briefly discussed in a preliminary report (*46*). It was suggested that when cholesterol is added, there is a partial release of sodium ions from the lamellae (*cf.* the linewidth experiment discussed above), giving rise to the initial decrease in the splitting. The increase in the quadrupole splitting of the sodium ions which occurs as the cholesterol content of the bilayer increases beyond 15 wt % may be explained by an increase in the order parameter, caused by a rearrangement of the molecular packing in the bilayer. Evi-

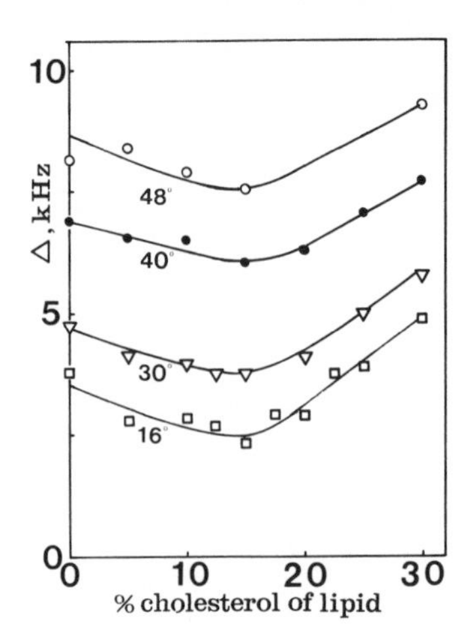

Figure 6. Observed ^{23}Na quadrupole splitting for lamellar mesophase samples of dimyristoyllecithin–cholesterol–$^{2}H_2O$ as a function of the cholesterol concentration at four different temperatures: 16°C (□), 30°C (▽), 40°C (●), and 48°C (○). In all samples the salt solution (0.8M NaCl in $^{2}H_2O$) accounted for 25% of the total sample weight. The concentration of cholesterol is expressed as the percentage (by weight) of cholesterol in the lecithin–cholesterol mixture.

dence from x-ray diffraction studies (kindly performed by K. Fontell) supports this idea. It was found (Figure 7) that at a molar ratio of 4:1 between lecithin and cholesterol (corresponding to a cholesterol concentration of 15% (w/w)) there is an abrupt change in the effect caused by the addition of cholesterol on the area occupied by each lipid molecule at the bilayer surface.

In the presence of cholesterol the splitting exhibits the same trend on both sides of the Chapman transition point for pure DML. This gives further support for the idea that cholesterol transforms the gel phase to

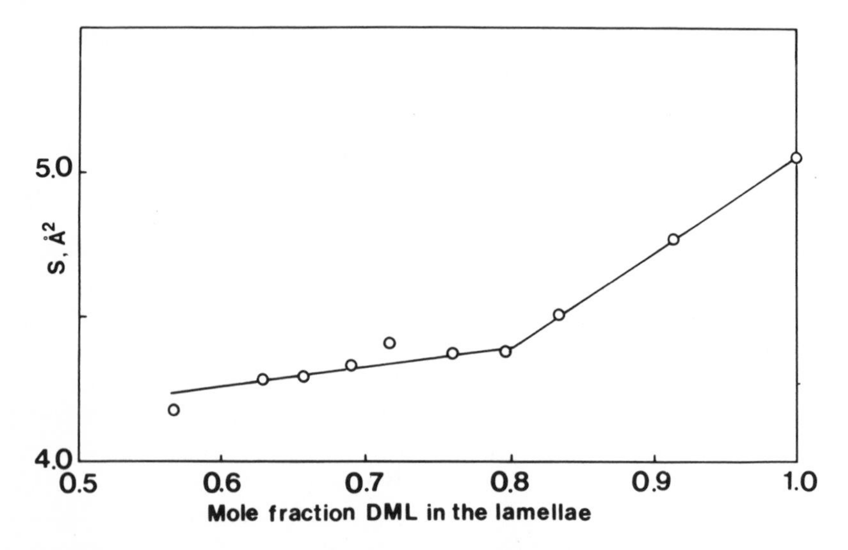

Figure 7. Area per molecule at the amphiphilic surface (calculated from x-ray data by K. Fontell) as a function of the mole fraction of dimyristoyllecithin (DML) in the lecithin–cholesterol bilayer

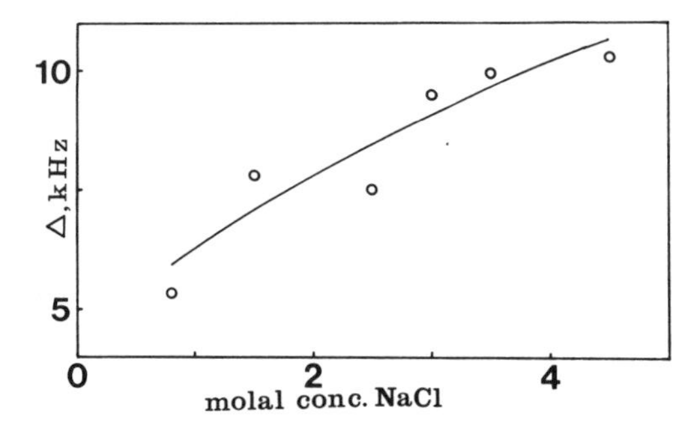

Figure 8. ^{23}Na quadrupole splittings as a function of the sodium chloride concentration in the water layer for lamellar mesophase samples composed of 75% dimyristoyllecithin and aqueous salt solutions accounting for 25% of the total weight

a liquid crystalline state (*47*). Recently we showed from a line shape analysis of proton magnetic resonance spectra of mesophases composed of lecithin and cholesterol that the effect of cholesterol is to enhance the phospholipid lateral diffusion in the bilayer (*48*).

Effect of Salt Concentration and Temperature. With increasing sodium chloride concentration there is a considerable increase in the ^{23}Na quadrupole splitting (Figure 8). This is also seen when the temperature of the sample is increased (*46*). However, if the number of bound sodium ions remained unchanged, one would expect a decrease in the splitting as the salt concentration or temperature is raised. We therefore suggest that increases in electrolyte concentration or temperature affect the conformation of the polar end groups of the lecithin molecules in such a way that the number of sodium ion binding sites is increased. This conclusion is supported by our deuteron NMR studies described below.

Deuteron NMR Studies. WATER DEUTERON QUADRUPOLE SPLITTINGS. Many water deuteron NMR studies of lyotropic mesophases have been reported (*10*), showing that detailed information on the water orientation can be obtained in this way. Mesophases composed of a soap and water have been studied extensively in regard to how the degree of orientation of water and alcohol OH groups depends on sample composition, soap end group, counterion, and temperature, for example (*19, 49*). Previously, we briefly reported water deuteron NMR studies of lecithin lamellar mesophases (*45, 46*) containing either egg-yolk lecithins or fully saturated synthetic lecithins. These investigations can be summarized as follows:

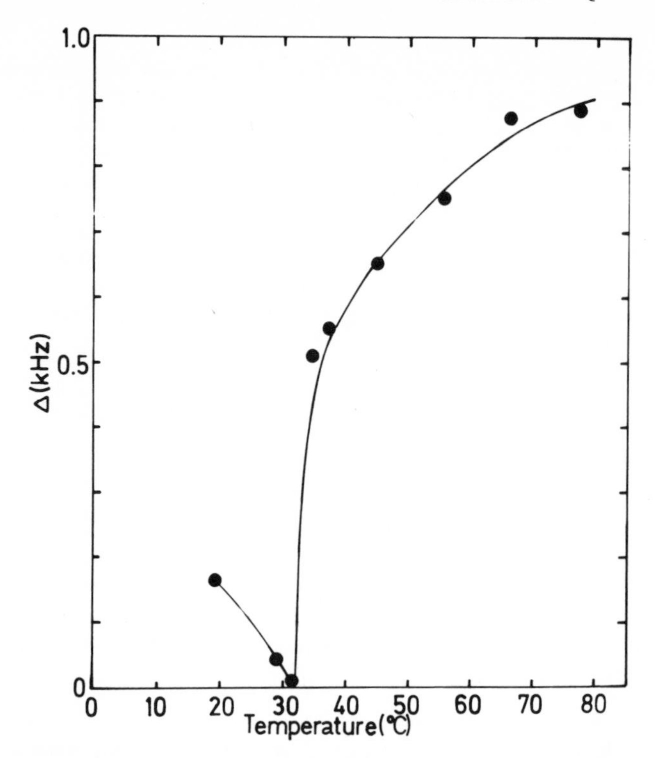

*Figure 9. Temperature dependence of the water deuteron quadrupole splitting of a lamellar mesophase sample composed of 75 wt % 1:1-mixture of dimyristoyl- and dipalmitoyllecithins and 25 wt % 0.8*M *NaCl in* ^2H_2O

(a) In analogy with soap–water systems (*49*) the water deuteron splitting decreases with increasing water content.

(b) Addition of cholesterol increases the deuteron splittings. This was interpreted as being caused by an intermediate exchange of deuterons between cholesterol-O^2H and ^2H_2O in analogy with previous findings in simple soap–water systems (*15*). However, the lecithin–cholesterol–water system seems more complex. Thus we have not observed separate deuteron NMR signals from the two deuteron sites in question, and we have not found any large changes in the splittings after changing the pH of the samples several units. (Deuteron exchange between O^2H groups and ^2H_2O is acid-base catalyzed and should therefore be markedly pH dependent.) Further studies are needed for an unambiguous interpretation of these deuteron NMR spectra.

(c) For egg yolk–lecithin systems the splitting was smaller with K^+ than with other alkali ions added to the lamellar mesophase samples (*cf.* ^{23}Na linewidth data discussed above).

(d) With increasing temperature there is an increase in the deuteron splitting for lecithin samples both with and without cholesterol. In Figure 9 the temperature dependence for a 1:1 mixture of dimyristoyl-

and dipalmitoyllecithin is shown. In all cases and in agreement with similar investigations by others (*50, 51*) we found that the splitting decreases with increasing temperature below the Chapman transition temperature whereas in the liquid crystalline region the splitting increases with increasing temperature. The effect of the added sodium chloride on the Chapman transition point for the different lecithins was negligible (*46*). It is interesting that near the transition temperature the deuteron quadrupole splitting approaches zero. At least two explanations can be offered for this phenomenon: (1) The water molecules are moving isotropically at the gel-to-liquid crystal transition. An isotropic environment might arise if, for example, the lamellae breaks down into smaller pieces at the transition temperature. This explanation seems, however, less probable since the temperature dependence of the ^{23}Na quadrupole splitting (*46*) and the ^{2}H quadrupole splitting of the choline–N–$C^{2}H_3$ group (*see* below) did not show any discontinuities at the phase transition temperature. (2) There is an exchange between sites having different signs on the order parameters. This could be understood if we consider the following simple model. Assume that only two binding sites are available for the water molecules—*viz.*, the phosphate and the choline groups. These two sites may have different signs on their order parameters (*cf.* calculations of the order parameter for different orientations of $^{2}H_2O$ molecules in a lamellar mesophase performed by Johansson and Drakenberg (*16*)). If the populations of water molecules are altered by temperature changes, the observed behavior of the splitting can be obtained. Such changes in the number of water molecules in the two sites could arise if there is a temperature-dependent interaction between the choline group and the phosphate group—*i.e.*, if the conformation of the choline group depends on the temperature. It should also be noted that the order parameter is zero when $\theta_{DM} = 54.7°$ (if $\eta = 0$), but since this has to happen exactly at the phase transition temperature, this is considered to be too unrealistic to be a probable explanation. We have started a systematic study of several different lyotropic mesophase systems with the hope of being able to solve this problem.

The increase in the ^{23}Na quadrupole splitting with increasing electrolyte concentration described in a previous section also led us to the suggestion that the conformation of the choline group is altered by salt addition. We assume that the interaction beween the $-N^{+}(CH_3)_3$ group and the phosphate group can be altered either through a specific interaction of the added ions with the charged groups of the lecithin molecules or by a non-specific effect arising from the change in ionic strength in the aqueous layers. By performing a competition experiment using the methods described above we hope to be able to decide which one of the effects is more important.

As shown in Figure 10 the water deuteron quadrupole splitting increases with increasing electrolyte concentration. This may well fit with the model discussed above if the postulated conformation change of the polar head groups consists of a stretching out of the choline groups away

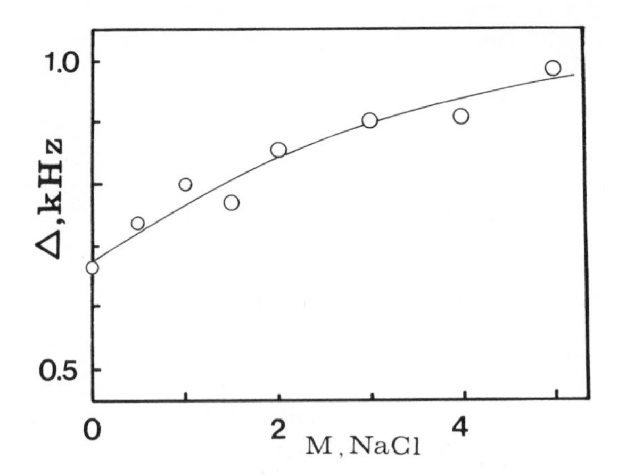

Figure 10. Water deuteron quadrupole splitting as a function of sodium chloride concentration for the same samples as in Figure 8

from the lamellar surface. Around every choline group we should then have an enhanced ordering of the water molecules, resulting in an increased deuteron quadrupole splitting.

DEUTERON QUADRUPOLE SPLITTINGS OF THE CHOLINE–N–C^2H_3 GROUP. The ^2H quadrupole splitting of the choline–N–C^2H_3 group of dimyristoyl lecithin was studied as a function of temperature (*see* Figure 11) and sodium chloride concentration. Contrary to what was observed for water ^2H splittings, no drastic change in the splitting is obtained at the Chapman transition temperature. A decrease in splitting with increasing temperature is expected since an increase in thermal mobility should cause a decrease in the order parameter (*cf.* Equation 8). From Equation 8 an order parameter, $S \simeq 0.01$, at 25°C has been calculated from the observed ^2H quadrupole splitting $\Delta_p = 1.5$ kHz. The quadrupole coupling constant (e^2qQ/h) was assumed to be 170 kHz, as determined for the C–^2H bond

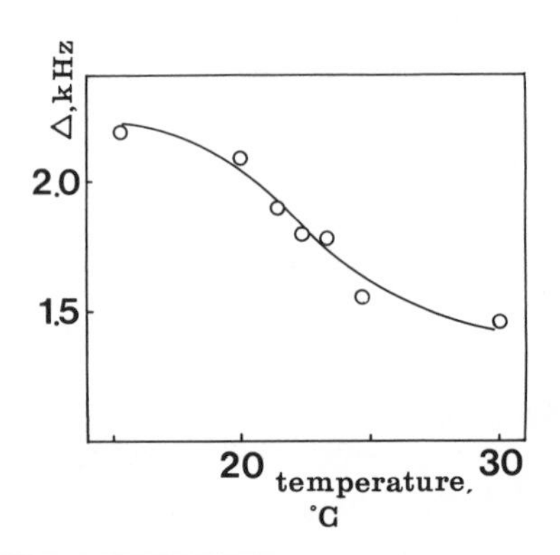

Figure 11. Temperature dependence of the choline–N–C^2H_3 deuteron quadrupole splitting of a lamellar mesophase sample composed of 75 wt % [N–C^2H_3] dimyristoyllecithin in H_2O

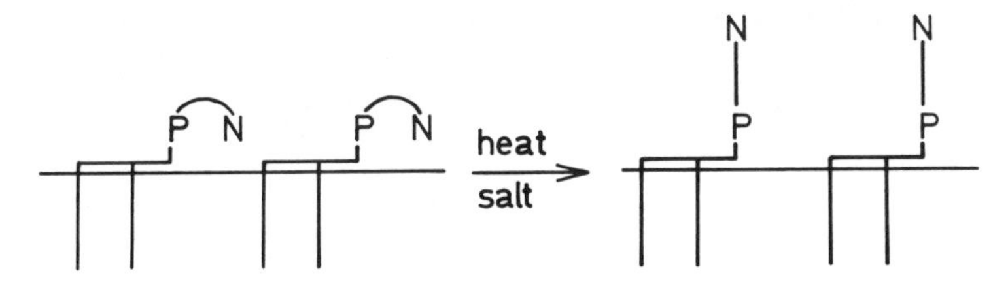

Figure 12. Schematic picture of suggested conformational changes of the phosphorylcholine group at the surface of the lecithin bilayers

in paraffins (*52*). A free rotation of the methyl group and an isotropic rotation about the choline–C–N(CH_3)$_3$ bond reduce the effective quadrupole coupling constant by a factor of about 9 (*see e.g.*, Ref. *53*). If these motions are taken into account, $S \approx 0.1$.

Conclusion

The data obtained so far on the effects of heat and electrolyte concentration on the quadrupole splittings of sodium ions and heavy water in lamellar mesophases of lecithins are all consistent with the model depicted in Figure 12. Although the proposed conformational change of the phosphorylcholine group still must be regarded as a tentative interpretation of our data, we have recently obtained additional results which strongly support the present model. Thus we found that the quadrupole splitting of the choline–N–C^2H_3 group is reduced when sodium chloride (3.0*M*) is added to the dimyristoyl lecithin sample. This certainly means that the addition of salt decreases the ordering of the choline groups as would be expected if they move out from the plane of the bilayer surface.

Future Aspects

The application of NMR spectroscopy in studies of ionic interactions and water structure in model membrane systems such as phospholipid lamellar mesophases is a relatively new field of research. It is, however, obvious from the results obtained that the presently described methods of NMR spectroscopy can provide unique information on membrane structure which cannot be gained by other methods. As soon as the results from model membrane systems can be interpreted unambiguously, these techniques should find meaningful applications in the study of biological membranes. Present experience leads us to expect rapid development in this direction. It may also be foreseen that investigations on biological membranes should be greatly facilitated by selective labelling *in vivo* of specific membrane components—*e.g.*, by feeding an experimental animal ^{2}H-labeled choline. In a recent preliminary deuteron NMR study (*54*) we have shown that this is possible in an investigation of rat-liver mitochondrial and microsomal membranes.

Literature Cited

1. Wilkins, M. H. F., Blaurock, A. E., Engelman, D. M., *Nature New Biol.* (1971) **230,** 72.
2. Steim, J. M., Tourtellote, M. E., Reinert, J. C., McElhaney, R. N., Rader, R. L., *Proc. Nat. Acad. Sci. U.S.* (1969) **63,** 104.
3. Hubbell, W. L., McConnell, H. M., *Proc. Nat. Acad. Sci. U.S.* (1969) **64,** 20.
4. Singer, S. J., Nicolson, G. L., *Science* (1972) **175,** 720.
5. Papahadjopoulos, D., in "Biological Horizons in Surface Science," L. M. Prince, D. F. Sears, Eds., p. 159, Academic Press, New York, 1973.
6. Chapman, D., in "Biological Membranes, Physical Fact and Function," D. Chapman, Ed., p. 125, Academic Press, New York, 1968.
7. Bangham, A. D., in "Progress in Biophysics and Molecular Biology," J. A. V. Butler, D. Noble, Eds., p. 29, Pergamon Press, Oxford, 1968.
8. Bangham, A. D., *Ann. Rev. Biochem.* (1972) **41,** 753.
9. Cohen, M. H., Reif, F., *Solid State Phys.* (1957) **5,** 321.
10. Johansson, A., Lindman, B., in "Liquid Crystals and Plastic Crystals," Vol. 2, G. W. Gray, P. A. Winsor, Eds., p. 192, Ellis Horwood Publishers, Chichester, 1974.
11. Saupe, A., *Angew. Chem.* (1968) **7,** 97.
12. Lindblom, G., Persson, N.-O., Lindman, B., in "Chemie, Physikalische Chemie und Anwendungstechnik der grenzflächenaktiven Stoffe," *Proc. Intern. Congr. Surface Active Substances, 6th, Zürich, 1972,* Vol. II, p. 939, Carl Hanser Verlag, München, 1973.
13. Wennerström, H., Lindblom, G., Lindman, B., *Chem. Scripta* (1974) **6,** 97.
14. Lindblom, G., *Acta Chem. Scand.* (1972) **26,** 1745.
15. Persson, N.-O., Wennerström, H., Lindman, B., *Acta Chem. Scand.* (1973) **27,** 1667.
16. Johansson, A., Drakenberg, T., *Mol. Cryst. Liquid Cryst.* (1971) **14,** 23.
17. Charvolin, J., Mannerville, P., Deloche, B., *Chem. Phys. Lett.* (1973) **23,** 365.
18. Seelig, J., Niederberger, W., *J. Amer. Chem. Soc.* (1974) **96,** 2069.
19. Wennerström, H., Persson, N.-O., Lindman, B., *ACS Symp. Ser.* (1975) **9,** 253.
20. Hubbard, P. S., *J. Chem. Phys.* (1970) **53,** 985.
21. Shporer, M., Civan, M. M., *Biochim. Biophys. Acta* (1974) **354,** 291.
22. Yeh, H. J. C., Brinley, F. J., Becker, E. D., *Biophys. J.* (1973) **13,** 56.
23. Cubero Robles, E., van den Berg, D., *Biochim. Biophys. Acta* (1969) **187,** 520.
24. Brockerhoff, H., Yurkowski, M., *Can. J. Biochem.* (1965) **43,** 1777.
25. Selinger, Z., Lapidot, Y., *J. Lipid Res.* (1966) **7,** 174.
26. Slotboom, A. J., de Haas, G. H., van Deenen, L. L. M., *Chem. Phys. Lipids* (1967) **1,** 317.
27. Stoffel, W., LeKim, D., Tschung, T. S., *Z. Physiol. Chem.* (1971) **352,** 1058.
28. Lindman, B., Ekwall, P., *Kolloid-Z. Z. Polym.* (1969) **234,** 1115.
29. Lindblom, G., Lindman, B., *Mol. Cryst. Liquid Cryst.* (1973) **22,** 45.
30. Persson, N.-O., Johansson, A., *Acta Chem. Scand.* (1971) **25,** 2118.
31. Woessner, D. E., Snowden, B. S., *J. Chem. Phys.* (1969) **50,** 1516.

32. Woessner, D. E., in "Mass Spectrometry and NMR Spectroscopy in Pesticide Chemistry," R. Haque, F. J. Biros, Eds., p. 279, Plenum Press, New York, 1974.
33. Charvolin, J., Rigny, P., *J. Phys.* (1969) **30,** C4-76.
34. Lawson, K. D., Flautt, T. J., *J. Phys. Chem.* (1968) **72,** 2066.
35. Charvolin, J., Rigny, P., *Chem. Phys. Lett.* (1973) **18,** 575.
36. Oldfield, E., Chapman, D., Derbyshire, W., *Chem. Phys. Lipids* (1972) **9,** 69.
37. Wennerström, H., Persson, N.-O., Lindman, B., *J. Magn. Res.* (1974) **13,** 348.
38. Fenichel, I. R., Horowitz, S. B., in "Biological Membranes," R. M. Dowben, Ed., p. 177, J. & A. Churchill, Ltd., London, 1969.
39. Hazlewood, C. F., *Ann. N.Y. Acad. Sci.* (1973) **204,** 1.
40. Lindblom, G., Lindman, B., in "Chemie, Physikalische Chemie und Anwendungstechnik der grenzflächenaktiven Stoffe," *Proc. Intern. Congr. Surface Active Substances, 6th, Zürich, 1972,* Vol. II, p. 925, Carl Hanser Verlag, München, 1973.
41. Gustavsson, H., Lindblom, G., Lindman, B., Persson, N.-O., Wennerström, H., in "Liquid Crystals and Ordered Fluids," J. F. Johnson, R. S. Porter, Eds., Vol. II, p. 161, Plenum Press, New York, 1974.
42. Lindblom, G., Lindman, B., Mandell, L., *J. Colloid Interface Sci.* (1973) **42,** 400.
43. Lindblom, G., Lindman, B., *Mol. Cryst. Liquid Cryst.* (1971) **14,** 49.
44. Lindblom, G., Persson, N.-O., unpublished data.
45. Persson, N.-O., Lindblom, G., Lindman, B., Arvidson, G., *Chem. Phys. Lipids* (1974) **12,** 261.
46. Lindblom, G., Persson, N.-O., Lindman, B., Arvidson, G., *Ber. Bunsenges. Phys. Chem.* (1974) **78,** 955.
47. Chapman, D., Williams, R. M., Ladbrooke, B. D., *Chem. Phys. Lipids* (1967) **1,** 445.
48. Ulmius, J., Wennerström, H., Lindblom, G., Arvidson, G., *Biochim. Biophys. Acta* (1975) **389,** 197.
49. Persson, N.-O., Lindman, B., *J. Phys. Chem.*, in press.
50. Salsbury, N. J., Darke, A., Chapman, D., *Chem. Phys. Lipids* (1972) **8,** 142.
51. Finer, E. G., Darke, A., *Chem. Phys. Lipids* (1974) **12,** 1.
52. Burnett, L. J., Muller, B. H., *J. Chem. Phys.* (1971) **55,** 5829.
53. Barnes, R. G., in "Advances in Nuclear Quadrupole Resonance," Vol. 1, J. A. S. Smith, Ed., p. 335, Heylen, London, 1974.
54. Arvidson, G., Lindblom, G., Drakenberg, T., unpublished data.

RECEIVED November 19, 1974. This work was supported by the Swedish Medical Research Council (Project No. 13X-4218) and the Swedish Natural Science Research Council.

10

Liquid Crystals and Cancer

E. J. AMBROSE

Chester Beatty Research Institute, Royal Cancer Hospital,
Fulham Rd., London SW3 6JB, England

A feature of most cancer cells is that they exhibit a disturbance of the ordered structures and biological regulation observed in normal cells. Evidence is presented, based on studies of living cells and on electron microscopy, which indicates that the arrangement of molecules within the plasma membrane and of subsurface microfilaments has certain properties that are similar to those of liquid crystals. These liquid crystalline properties are disturbed in malignant cells.

Cancer biologists have recognized for a number of years that one of the most characteristic features of cancer cells is a deviation from the organized and coordinated behavior of normal cells, but it has been extremely difficult to pinpoint these studies in molecular terms. With the finding of a more precise molecular basis for describing the structure of biological membranes and the molecular aggregates known as cytoplasmic filaments which occur within the cell cytoplasm, a description of certain aspects of cell behavior that are observed in malignancy has become possible.

In physicochemical terms, liquid crystals are usually defined as a state of matter of a single molecular species. In living organisms, numerous types of both small and large molecules occur. Nevertheless, certain properties of these multicomponent systems can be recognized as being related to properties that are studied by physical chemists when they are investigating liquid crystals. In cancer cells, the three cellular components which are of particular interest are the phospholipids of the plasma membrane, the actinlike subunits of the microfilaments which lie just beneath the plasma membrane, and the microtubules which lie deeper in the cytoplasm. A particularly interesting approach to the general problem of evidence for decreased order within the cellular structures of cancer cells has been NMR studies which provide evidence

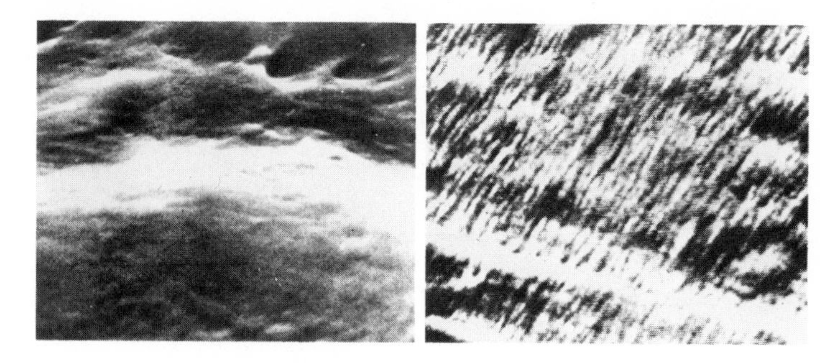

Figure 1. Fibroblast cells as seen by stereoscan microscopy
Left: Surface of cell in tissue culture (× 18,000)
Right: Cell after removal of the plasma membrane by ion etching to reveal oriented subsurface microfilaments (similar to Figure 2c) (× 18,000)

for increased molecular tumbling of the water molecules in the cytoplasm (*1*). The relationship between these various studies that indicate a disturbance of molecular organization in cancer cells is considered in this paper.

Subsurface Structures with the Appearance of Nematic Liquid Crystals

Beneath the outer or plasma membrane of mammalian cells lies a subsurface region of the cytoplasm. A simple way to observe this subsurface region is to remove the plasma membrane by etching and then to examine it by stereoscan microscopy (*2*). Although liquid crystals in general are highly labile, the large scale structures that occur in living cells can be reasonably well preserved by suitable fixation, particularly by freeze substitution or by the use of glutaraldehyde followed by freeze drying or freeze substitutions (*3*). The plasma membrane can be removed by bombarding the specimen with hydrogen ions in a high frequency electrical discharge (ion etching). For the results of such an experiment, *see* Figure 1 which pictures the outer surface of the plasma membrane (Figure 1, left) and the subsurface structure after etching (Figure 1, right). From studies by standard transmission electron microscopy, together with the findings from various specific reactions with heavy meromyosin, it is now clear that these subsurface microfilaments (diameter ~ 60 A) are very similar to the actin filaments of smooth muscle. With these are associated myosin molecules; the actin–myosin-like complex has calcium-sensitive, magnesium-dependent ATPase activity like that of smooth muscle.

Microfilaments of the actin type are formed by linear aggregation of corpuscular molecules of ~ 60 A diameter (Figure 2e). The filaments

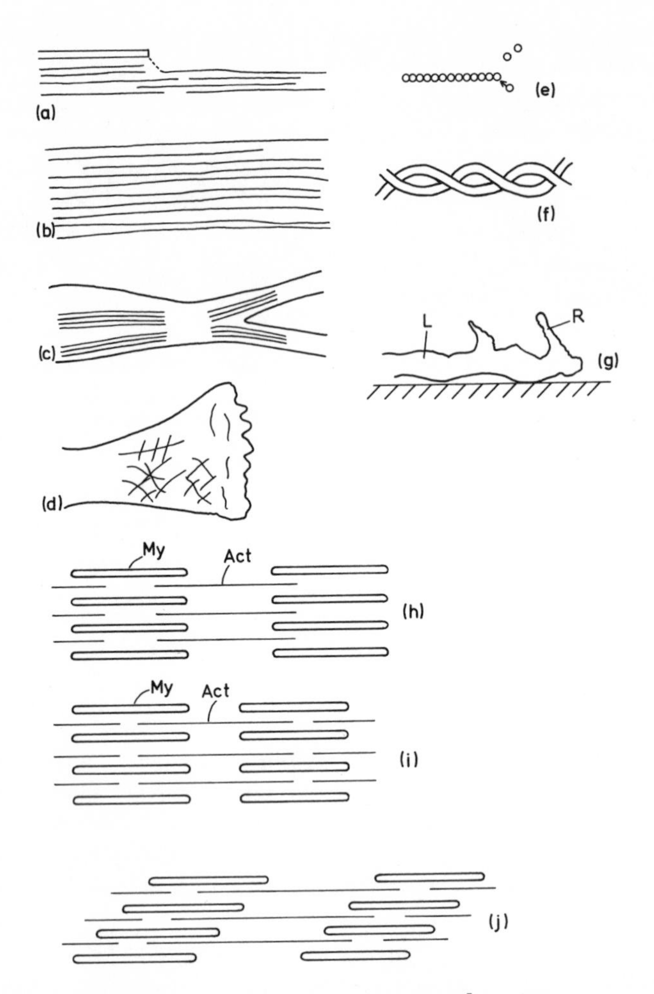

Figure 2. Cell structure in mammalian tissue

(a) Section through the surface region of a cell showing subsurface microfilaments exposed by etching away part of the surface with hydrogen-ion bombardment
(b) Plan of subsurface region of a cell bounded by plasma membrane. Microfilaments are aligned as in Figure 1b.
(c) Bundles of microfilaments (cortical microfilaments), some lying in branching pseudopodia
(d) Network microfilaments in the leading edge region of a migrating fibroblast (lamellar region)
(e) Formation of microfilaments by aggregation of subunits
(f) Coiled filaments in the actin of smooth muscle
(g) Section through the leading edge of a migrating fibroblast moving on a glass surface (L: lamellar cytoplasm, R: ruffled membrane)
(h) Structure of striated muscle before contraction (schematic) (My: myosin, Act: actin)
(i) Structure of striated muscle after contraction. Actin filaments slide toward each other between myosin molecules. Myosin head proteins attach to the actin filaments (not shown).
(j) Staggered arrangement of myosin and actin in smooth muscle. During contraction, both actin filaments and myosin molecules slide past each other.

coil in twos into helices (Figure 2f). It is not known why the filaments form in the subsurface region of the cytoplasm. This is the region of various gradients (*e.g.* ionic gradients and electrical potentials) close to the membrane, which may possibly affect the aggregation of the subunits. There are various ways in which the filaments may be associated. Since they are near the boundary provided by the cell membrane, there will be a marked tendency to produce a molecular felt (Figure 2a). When there is a containing sheet or tube of plasma membrane, this may assist in the alignment of the molecular felt as in Figures 1, left and 2b. It is clear from stereoscan and transmission electron microscopic studies that the filaments tend to associate in bundles (cortical microfilaments) (Figure 2c). These are particularly conspicuous in glial cells which produce extremely long and sometimes branching pseudopodia; it is possible that these bundles, together with the microtubules described below, may play some role in maintaining these elongated pseudopodia as is suggested in Figure 2c.

An interesting relationship exists between the actinomyosinlike subsurface microfilaments found in fibroblasts, epithelial cells, glial cells, etc., and the microfilaments of smooth and striated muscle. Figure 2h depicts the relationship between actin filaments and myosin filaments in striated muscle. The myosin molecules do in fact have a massive head protein (heavy meromyosin) which contains the ATPase activity and associates with the actin filaments during contraction. According to the Huxley and Hanson (*4*) sliding filament model, muscular contraction takes place as depicted in Figure 2i. As it occurs in the limbs of mammals, this type of muscular contraction is rapid and limited in the extent of contraction. It is a highly coordinated activity controlled by motor neurons. During the contraction of striated muscle, the actin filaments slide within the myosin assemblies, as shown.

In smooth muscle, the alignment of myosin molecules is less precise; the neighboring molecules may be staggered as in Figure 2j. During contraction, the actin filaments slide with respect to myosin molecules, and myosin molecules slide with respect to each other. This gives possibilities for much greater contraction, and a state of contraction can be maintained for long periods, *e.g.* in clam muscle it may persist for several days.

With the subsurface filaments, there is a further progression toward a lower degree of ordering of actinomyosinlike contractile elements. With the filaments in Figure 1b, a linear contraction rather similar to that of smooth muscle might be expected whereas in regions of network filaments (Figure 2d), it might be expected that the contractions would be less coordinated, approximating somewhat the supercontraction of biochemical preparations of actinomyosin gels when ATP is added.

Cellular Function of Microfilaments

Clearly, in smooth and striated muscle, the actinomyosin filaments are synthesized to such an extent that the cytoplasm becomes filled with filaments; the orientation along the elongated muscle cell is high and the structure in Figure 2h resembles somewhat that of a smectic phase of a lyotropic liquid crystal. The biological effect is to produce linear contraction by sliding of the filaments. In smooth muscle, the appearance is somewhat similar to that of a nematic phase, again producing linear contraction. In other cell types such as fibroblasts and epithelial cells, microfilament synthesis is much reduced. The subsurface sheets (Figure 2b) now play a role in generating cell surface movements. For example, contraction of such a subsurface sheet may produce wrinkles in the plasma membrane, thereby initiating the formation of surface waves (undulating membranes) like sea waves (Figure 2g). These waves play an important role in cellular locomotion, enabling the cell to progress along the surface by making intermittent contact with the substratum (Figure 2g).

On the other hand, the network filaments (Figure 2d) would be expected to produce more irregular surface movements. They are found particularly on the leading and spreading edges of the cell (Figures 2d and 2g). They produce large scale ruffled membranes which seem to attach to the substratum and to provide the main locomotive force by their subsequent contraction. Evidence that the microfilaments are closely involved in the cellular locomotion of tissue cells has come from use of the drug cytochalasin B. This drug causes a reversible disturbance of the microfilament structure. These changes are visible in electron microscope sections (*5*). When treated with this drug, cells are unable to generate surface movements that are capable of initiating cellular locomtion. When the drug is removed, the cells regain their ability to produce cell surface movements, and they move on the substratum.

Changes in Subsurface Microfilaments in Cancer

In malignant cells, the organization of subsurface microfilaments is reduced (*2, 6*). This change is apparent when malignant brain tumor cells are compared with the normal human glial cells from which they originated. Similar conclusions were reached by McNutt, Culp, and Black (*7*) who used replica and transmission electron microscopic methods, and by Vasiliev and Gelfand (*8*) who used stereoscan as well as transmission electron microscopy. There is a general progression from the normal cell to the more differentiated tumor cell and finally to the highly anaplastic and invasive malignant cell (Figures 3f and 3g). Some orientation of filaments is preserved in the more differentiated cells, but they start to form on their leading edge irregular pseudopodia that are variable in length and diameter. The regular formation of wavelike movements is reduced. These irregular pseudopodia have been called polypodia. They are associated with surface regions where no organized subsurface filaments (cortical microfilaments) can be observed. In the highly malignant cells, the coordination of cell surface movements is lost, and highly irregular polypodial movements are produced.

Cytoplasmic Microtubules

Cytoplasmic microtubules represent another type of cellular fiber that is well known as forming the mitotic spindle during mitosis. They are also present in the cytoplasm of cells during interphase. Like microfilaments, they are formed by the aggregation of corpuscular units into filaments (Figure 3c), but in this case the filaments then aggregate further to form hollow tubes 270 A in diameter (Figure 3b). Because of their tubular form, these fibers are comparatively rigid. They also aggregate to form structures that resemble somewhat nematic liquid crystals. They grow from an aggregation center, generally arising from bodies known as centrioles which lie close to the nuclear membrane in animal cells. They therefore tend to lie mainly within the internal cytoplasm. As with actinomyosin, when microtubules are synthesized in large quantities so that they fill the cytoplasm, they can produce highly ordered structures. For example, axonemes of heliozoa are packed with microtubules which in transverse section lie in a spiral array (Figure 3d) (*9*).

Biological Function of Microtubules

Microtubules can be disrupted by treating cells with colcemid, without damaging the locomotory function of the cell. Such treatment of elongated cells such as glial cells in culture causes them to lose their spiky pseudopodia (Figure 3e) (*5*). Nevertheless they are still able to migrate on the substratum. They show undulating cell surface movements as before; the subsurface microfilaments are not disturbed. This and other evidence indicates that the microtubular assemblies play a role in providing a cytoskeleton. They enable the cell to maintain an elongated form, particularly in the case of very long and spikelike pseudopodia.

As far as cancer is concerned, disturbances of microtubular organization do not appear to play an impotrant role (*10, 11, 12*). Disturbances of these structures by colcemid and loss of long spiky pseudopodia do not affect the capacity of cancer cells to invade normal tissues.

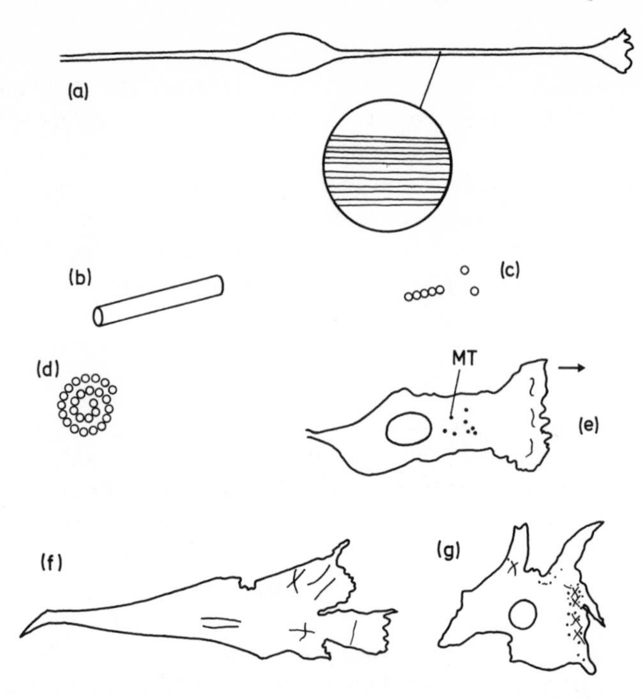

Figure 3. Structure of normal and malignant cells

(a) Elongated glial cell (connective tissue cell of the brain). Inset shows microtubules lying in the elongated pseudopodia. (b) Shape of a microtubule (diameter, 270 A)

(c) Formation of filaments constituting microtubules from subunits

(d) Transverse section through the spikey pseudopodium (anoneme) of a heliozoan showing spiral packing of microtubules (e) Effect of treatment with colcemid (5–10 μg/ml) on the shape of the glial cell in (a). Microtubules break down, but ruffled membranes are generated on the leading edge and the cell continues to move.

(f) Moderately malignant tumor cell derived from a fibroblast. Some breakdown of subsurface microfilament organization has occurred. Irregular pseudopodia (polypodia) are produced on the leading edge.

(g) Anaplastic tumor cell showing irregular surface movements of polypodia and complete loss of organization of subsurface microfilaments

Liquid Crystalline State of the Plasma Membrane

The Singer and Nicholson (*13*) model for the plasma membrane, which now receives much support, is basically a smectic liquid crystal consisting of one bilayer of phospholipid (Figure 4a). The phospholipid bilayer contains cholesterol at a concentration which depends on cell type. Embedded in the lipid liquid crystal lie protein molecules. Some of these protein molecules transverse the entire lipid bilayer and communicate both with the inside and the outside of the cells. Some of these may

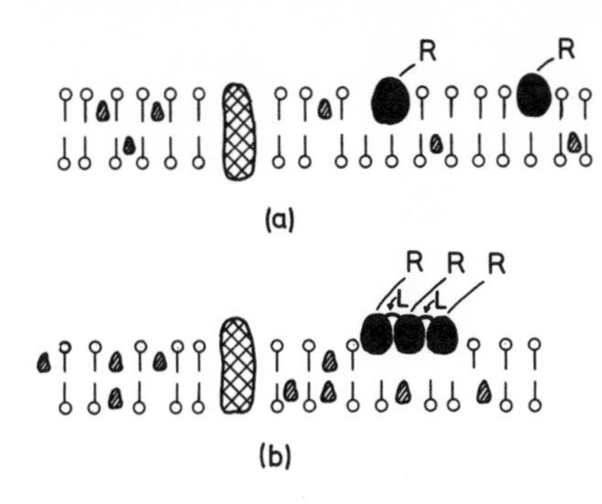

Figure 4. Structure of the Singer and Nicholson fluid mosaic model for the plasma membrane

(a) Section of normal cell (schematic). The phospholipid bilayer contains cholesterol molecules; protein molecules are dispersed within the liquid crystal bilayer (R: receptor for wheat germ agglutinin)

(b) Section of tumor cell showing increased cholesterol content of the lipid bilayer and aggregation of wheat germ receptors

be involved in transport phenomena. Other proteins are only partially embedded in the phospholipid bilayer. This fluid mozaic model allows the protein molecules considerable motility. Certain of these proteins contain receptor sites for proteins obtained from plant seed germ, known as lectins, which are capable of agglutinating cells.

Of particular interest is a protein obtained from wheat germ which has specificity for cancer cells (*14*). A purified protein (*15*) that was prepared from wheat germ specifically agglutinates a suspension of malignant cells in preference to the normal cells from which they were derived. The work of Singer and Nicholson (*13*) indicates that this change in agglutinability of cancer cells is attributable to an alteration in the position of the receptor sites which come together and form clusters (Figure 4b). Whether this change in the fluid mozaic is caused by an alteration in the liquid crystalline properties of the lipid phase (*e.g.* most tumor cells have an increased cholesterol content in the membrane) or whether it is caused by an increased affinity of receptor site proteins for each other is not known at present. Nevertheless, these phenomena, as depicted in Figure 4, clearly illustrate the importance of liquid crystalline phenomena in explaining the surface properties of malignant cells.

General Conclusions Concerning the Involvement of Liquid Crystalline Phenomena in Cancer

The importance of cell surface changes in malignancy has long been recognized (*16*), but until recently studies in this field were concerned mainly with the peripheral surface properties and with glycoproteins, particularly the acidic sialomucoproteins. Reduced cell–cell adhesiveness is one of the most striking biological characteristics of cancer cells. For example, the sheet of normal epithelial cells in tissue culture (Figure 5a) has continuous cell contacts at the cell borders. The earliest change observed in malignancy is the appearance of gaps between the cells which indicates reduced cell–cell adhesiveness (Figure 5b). Such

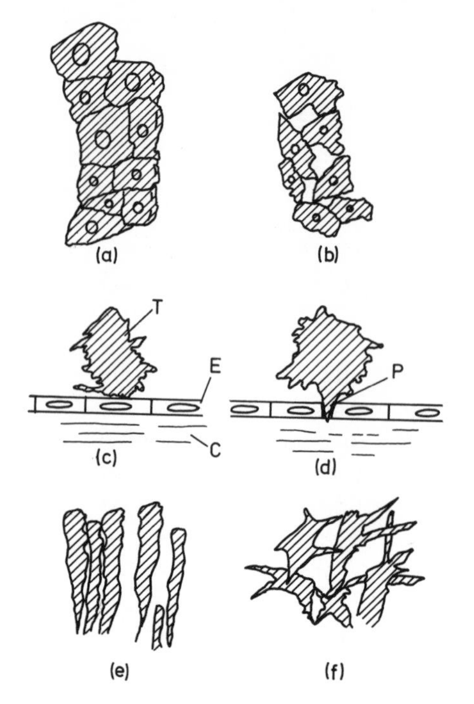

Figure 5. Structure of normal and malignant tissue
(a) Sheet of normal epithelial cells in tissue culture showing continuous contact of the cell borders
(b) Moderately differentiated carcinoma cells derived from epithelium showing gaps in cell–cell contacts
(c) Section through a normal membranous tissue (chick chorioallantoic membrane) (E: outer layer of epithelium, C: connective tissue, T: tumor cell showing polypodial activity when it is first placed on the normal tissue)
(d) The same as (c), 2 hrs later. A polypodial extension of the tumor cell has forced a passage between the normal epithelial cells thereby initiating the process of tumor invasion.
(e) Group of normal fibroblasts in monolayer tissue culture in normal culture medium with normal physiological potassium concentration (6mM)
(f) The same as (e), in the presence of 35mM potassium. The normal cells have acquired the morphology and contact behavior of tumor cells (this process is reversible).

changes in the biochemical constitution of the surface may involve conformational and packing phenomena of the glycoproteins in which the liquid crystalline properties of the fluid mozaic membrane (described above) may play a part. However, from this point of view, it is principally the disturbance of the liquid crystalline order of the subsurface microfilaments which is now attracting interest in cancer research. Perhaps these changes should be regarded as symptomatic of the general tendency toward a reduction in the molecular order in cancer tissues, changes which are not yet fully explainable in terms of liquid crystalline phenomena.

Nevertheless, the involvement of the surface polypodia in the characteristic biological properties of cancer cells has now been demonstrated. Even before noticeable changes in cell–cell contacts can be observed, the appearance of ragged leading membranes of cells, showing loss of subsurface microfilament organization, may be indicative of invasive cancer. The importance of the polypodia so formed is illustrated by time-lapse pictures of the contacts between malignant cells and whole tissues (*11, 12*) (Figures 5c and 5d). Figure 5c depicts a section of a natural membranous tissue with an epithelial sheet on its outer surface. When a cancer cell is first placed on the sheet, it exhibits irregular polypodial surface movements, but, quite rapidly, one of these surface probes penetrates between the normal cells and secures a point of anchorage (Figure 5d), thereby initiating the invasive process.

As to the origin of the alterations in molecular ordering of subsurface microfilaments, little is known at present, but it is not impossible that changes in the ionic milieu could account for these changes in the state of aggregation. We have found that a change in the external potassium concentration surrounding normal fibroblasts does not affect their capacity to grow, divide, and migrate, but it does cause them to lose the smooth undulating membranes of their surface and to generate irregular polypodial movements like cancer cells (Figures 5e and 5f). This finding suggests that a change in the permeability of the membrane itself that leads to a changed ionic milieu of the cytoplasm could be a factor in disturbing the supermolecular structure of the cytoplasm. Additional experimental work is needed before these characteristics of cancer cells can be fully understood.

Literature Cited

1. Damadian, R., *Science* (1971) **171,** 1151–1153.
2. Ambrose, E. J., Batzdorf, U., Osborn, J. S., Stuart, P. R., *Nature London* (1970) **227**, 397–398.
3. Ambrose, E. J., Ellison, M. L., *Eur. J. Cancer* (1968) **4,** 459–462.
4. Huxley, H. E., Hanson, J., *Nature London* (1954) **173,** 973.
5. Wessells, N. K., Spooner, B. S., Ludueña, M. A., *CIBA Found. Symp.* (1973) 53–76.
6. Ambrose, E. J., Batzdorf, U., Easty, D. M., *J. Neuropathol. Exp. Neurol.* (1973) **31,** 596–610.
7. McNutt, N. S., Culp, L. A., Black, P. H., *J. Cell Biol.* (1971) **50,** 691–708.
8. Vasiliev, J. M., Gelfand, I. M., *CIBA Found. Symp.* (1973) 309–328.
9. Porter, K. R., *CIBA Found. Symp.* (1966) 308–343.
10. Easty, D. M., Easty, G. C., *Br. J. Cancer* (1974) **29,** 36.
11. Ambrose, E. J., Easty, D. M., *Differentiation* (1973) **I,** 39–50.
12. *Ibid.* (1973) **I,** 277–284.
13. Singer, S. I., Nicholson, G. L., *Science* (1962) **175,** 720–731.
14. Ambrose, E. J., Dudgeon, J. A., Easty, D. M., Easty, G. C., *Exp. Cell Res.* (1961) **24,** 220–227.
15. Burger, M. M., Goldberg, A. R., *Proc. Natl. Acad. Sci. U.S.A.* (1967) **57,** 359–366.
16. Abercrombie, M., Ambrose, E. J., *Cancer Res.* (1962) **22,** 104–110.

RECEIVED November 19, 1974.

INDEX

The text of this book is set in 10 point Caledonia with two points of leading. The chapter numerals are set in 30 point Garamond; the chapter titles are set in 18 point Garamond Bold.

The book is printed offset on White Decision Opaque Regular, 60-pound. The cover is Joanna Book Binding blue linen.

Jacket design by John Sinnett.
Editing by Mary Rakow.
Production by Joan Comstock.

The book was composed by the Mills-Frizell-Evans Co. and by Service Composition Co., Baltimore, Md., printed and bound by The Maple Press Co., York, Pa.